ENVIRONMENTAL MANAGEMENT FOR SUSTAINABLE DEVELOPMENT

By

Prof. (Dr.) Rakesh Kumar Yadav
Pro VC & Officiating Vice Chancellor
Shri Venkateshwara University, Gajraula
Uttar Pradesh (India)

Dr. Shivani Kaul
Associate Professor
GNIOT Institute of Technology & Management
Greater Noida, Uttar Pradesh (India)
&

Ms. Swati Singh
Associate Professor
Department of Management
Shri Venkateshwara University, Gajraula
Uttar Pradesh, (India)

DPH

DISCOVERY PUBLISHING HOUSE
INDIA

Published by:

DISCOVERY PUBLISHING HOUSE
4383/4B, Ansari Road, Darya Ganj
New Delhi-110 002 (India)
Phone : +91-11-23279245; 23253475; 43596065
E-mail : discoverybooksindia@gmail.com
discoverypublishinghouse@gmail.com
namitwasan9@gmail.com
web : www.discoverypublishinggroup.com

First Edition: **2023**

ISBN: 978-81-958210-4-4

Environmental Management for Sustainable Development

Printed at:
Infinity Imaging Systems
Delhi

PREFACE

For young researchers from various institutions and colleges, this book has been structured in accordance with the most recent developments in environmental management.

The live and let live concept is an old one, but the modern tendency in science has embraced it as well as the extremely modern instruments and techniques needed to determine the origins and effects of the continuing developments in science that humankind had not previously understood.

The current book covers all pertinent research and current developments in environmental management.

With the most recent knowledge on many scientific topics, we are confident that new edition of the book will be more helpful to all researchers who have been active in scientific study.

The authors also like to express their sincere gratitude to all the academics, researchers, writers and scientists whose work has been referenced or otherwise incorporated into various portions of the book. Finally, we would like to express our gratitude to the publishers for taking an interest in the subject and for creating the book.

–Authors

CONTENTS

CHAPTER 1

Environmental Pollution

Environmental pollution is now a household word that calls up images of despoiled water, air, or land. Pollution eludes simple definition, because it may involve hundreds of factors that arouse from numerous sources. Some changes, such as contamination of drinking water, air that we inhale and food material that we take regularly, may directly affect human health and well being while others may affect humans in a more indirect way, such as carbon dioxide emissions affecting the climate, which in turn may affect food production, or change in the nutrient levels causing some population to die out and others to explode. Thus pollution may be defined as "the **unfavourable alteration of our surroundings, wholly or largely as a by product of man's action**." Pollution may also be defined as "**Addition of any foreign material (inorganic, biological or radiological) or any physical change in the natural environment which may adversely affect the living resources directly or indirectly, immediately or after sometime or after a very long time."** One definition of pollution is "An undesirable change in the physical, chemical or biological characteristics of air, water and soil that may harmfully affect the life or create a potential health hazard for any living organism". Pollution is thus direct or indirect change in any component of biosphere that is harmful to the living component (s), and in particular undesirable for man, affecting adversely the industrial progress, cultural and natural assets or general environment.

Environmental pollution has assumed a global magnitude and its frontiers are no more confined to any particular part of our planet earth. The first ever UN conference on environment held in Stockholm on 5th and 6th June 1972, known as United Nations Conference on Environment and Development (UNCED) had been able to focus the attention of world community on the growing world wide menace of environmental pollution.

Every organism in a natural ecosystem produces potentially polluting waste products. What makes natural ecosystem sustainable is that the waste from one kind of organism becomes food and or raw material for another. In balanced ecosystems, waste do not accumulate to produce unfavourable alterations, they are broken down and recycled. Through much of their history humans have relied on the same natural process to dispose of their wastes. But the situation has become extremely unbalanced. Exploding human population coupled with increasing use of materials and energy have led to enormous volumes of wastes and other materials being discharged in to the environment. Even when materials are biodegradable, that is, of a kind that can be assimilated and recycled by organisms, sheer volumes overwhelm the capacity of natural system to cope. Aggravating the problem is the production of increasing amounts and kinds of non-biodegradable materials, which are not readily broken down and assimilated by the natural processes.

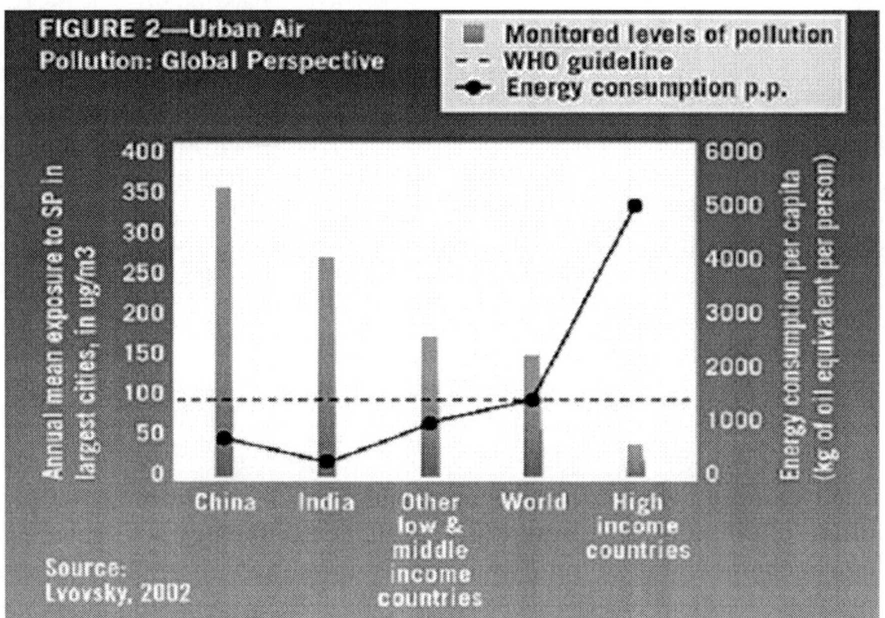

India today is one of the first ten industrialized countries of the world. Today we have a good industrial infrastructure in core industries like metals, chemicals, fertilizer, petroleum, food etc. What has come out of these? Pesticides, detergents, plastics, solvents, fuels, paints, dyes, food additives etc. are some examples. Due to progress in atomic energy, there has also been an increase in radioactivity in the biosphere. Besides these, there are a number of industrial effluents and emissions particularly poisonous gases in the atmosphere. Mining activities has also added to this problem particularly as solid waste.

Clearly, pollution involves so many different factors from so many sources that there is no single or simple remedy. Pollution is an evil which has born out of development. Lack of development of a culture of pollution control, has resulted a heavy backlog of gaseous, liquid and solid pollution in our country. It is to be cleaned. Thus pollution control in our country is a recent environmental concern. Not only in India, but also in the undeveloped Western World, pollution is a scare-word, which is a man-made problem, mainly of affluent countries. The developed countries have been in a mad race to exploit every bit of natural resource to convert them into goods, for their comfort, and to export them to needy developing world. In doing so, the industrialized countries dump lot of materials in their environment which becomes polluted. In one way pollution has been in fact "exported" to developing countries.

The following reports give the status of our air, water and soil which is so much so polluted that immediate steps need to control the situation: About 7000 tonns of solid waste is generated in Mumbai, the metro also generates about 3000 million liters of sewage every day, half of which is not treated. In water samples tested in 2003, 19% were found unfit for drinking. The Sulpher-dioxide levels in the metro exceeds 24 hours standards by 2 to 8%, nitrogen oxide by 17 to 56%, SPM by 31 to 83% and lead by 1 to 3%. The noise pollution of the metro exceeds all standards.

Similarly, Centre for Science and Environment (CSE) have reported harmful pesticides like lindane, DDT, Chloropyrifos and melathion in popular brands of bottled water in Delhi and Mumbai. In one of the brand, pesticide residues were 104 times higher than the prescribed norms of Europian Union. These pesticides get accumulated in the human body and cause cancer, nervous disorders and weaken the immune system.

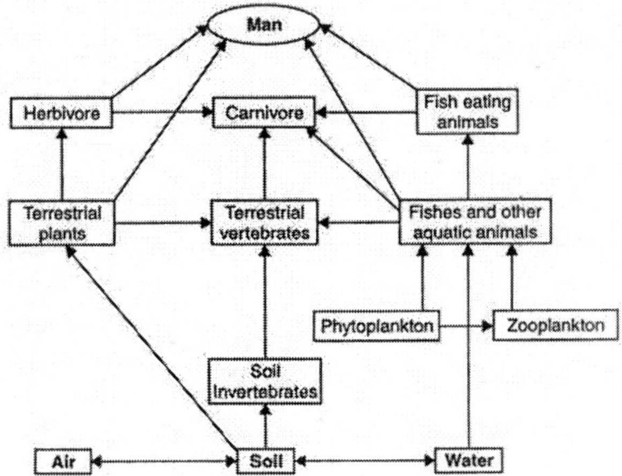

Fig. 1.1 : Path of different chemicals from air, soil and water to man

ENVIRONMENTAL POLLUTANTS

Substance which are responsible for a change in the natural environment or conditions, which are harmful to the nature or particularly to the human beings are called pollutants. A pollutant may thus include any chemical or geo-chemical (dust, sediment, grit etc.) substance, biotic component or its product, or physical factor (heat) that is released intentionally or unintentionally by man into the environment in such a concentration that may have adverse harmful or unpleasant effects. A pollutant has also been defined as "any solid, liquid or gaseous substance present in such concentration which may be or may tend to be injurious to the environment". Pollutants are the residues of things we make, use and throw away. There are many sources of such pollutants. The lakes and rivers are polluted by wastes from chemical and other factories, and the air by gases of automobile exhausts, industries, thermal power plants etc. As the clean air moves across the earth surface, it collects the products of both natural events such as dust storms, volcanic eruptions etc. and human activities such as vehicular emissions etc. These potential pollutants, called **primary pollutants**, mix with the churning air in the troposphere. Some may react with one another or with the basic components of air to form new pollutants called **secondary pollutants**. Long lived pollutants can travel great distances before they return to the earth surface as solid particles, droplets, or chemicals dissolved in precipitation.

The following pollutants have been identified as most wide spread and serious:

(*i*) **Particulate:** Particulate are tiny solid or liquid particles suspended in the air. These particles can be seen as smoke or haze. Other pollutants present as gas or vapor are not visible except in the case of nitrogen oxide, which is a brownish gas. Particulate may carry any or all of the other pollutants dissolved in or adhering to their surfaces.

(*ii*) **Hydrocarbons and other volatile organic compounds**: These include materials such as gasoline, paint solvents, and organic cleaning solutions that evaporate and enter the air in a vapor state.

(*iii*) **Carbon monoxide**: This is a highly poisonous gas

(*iv*) **Nitrogen oxides**: These include several nitrogen -oxygen compounds, all in gaseous forms.

(*v*) **Sulfur oxides:** Sulfur di-oxide is a poisonous gas to both plants and animals.

(*vi*) Lead and other heavy metals.

(*vii*) Ozone and other photochemical oxidants: Ground level ozone is a serious pollutant.

(*viii*) Acid droplets.

(*ix*) **Metals:** Mercury, lead, iron, zinc, nickel, tin, cadmium, chromium etc.

(*x*) **Agrochemicals:** Pesticides, herbicides, fungicides, nematicides bactericides, weedicides and fertilizers etc.

(*xi*) Various effluents and solid wastes which are not treated prior to their discharge.

(*xii*) Radioactive waste.

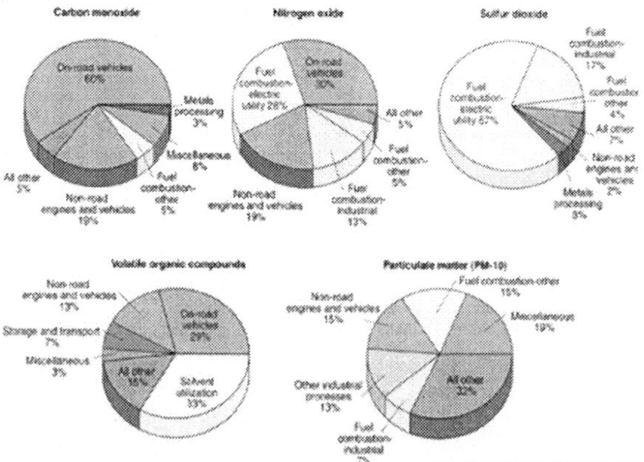

Fig. 1.2 : Different pollutants and their sources

On the basis of their being bio-degradable or non- degradable nature the pollutants may be grouped in one of the following two categories:

1. **Non-degradable pollutants :** These are the materials and poisonous substances like aluminium cans, mercuric salts, long-chain phenolics, DDT etc. that either do not degrade or degrade only very slowly in nature. They are not cycled in ecosystem naturally. They not only accumulate but also are often biologically magnified with their subsequent movement in food chains and biogeochemical cycles.

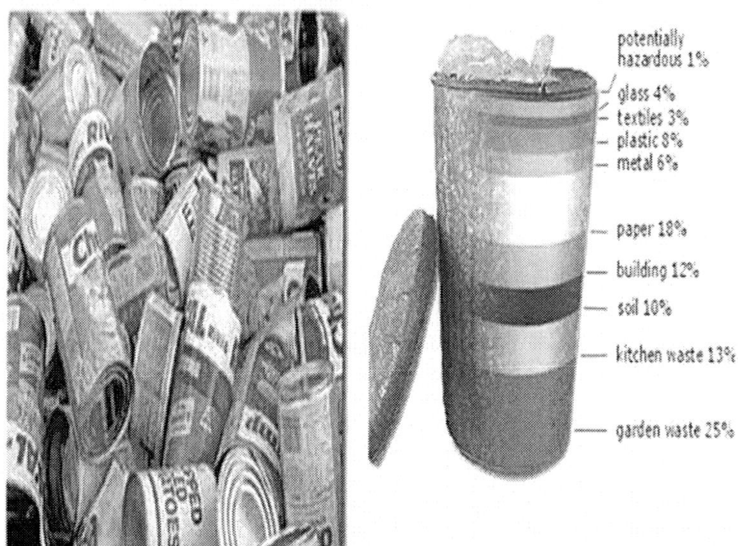

Fig. 1.3 : Different types of non-biodegradable and bio-degradable pollutants

2. **Biodegradable pollutants :** They are the domestic wastes than can be rapidly decomposed under natural conditions. They may create problems when they accumulate (*i.e.* their input into the environment exceeds their decomposition).

In United Nations Environment Program document following order of priority of different pollutants has been indicated:

S. No.	Order of priority	Medium
1.	SO_2 + suspended particle, strontium, cesium.	Air and food
2.	Ozone DDT and other organochlorine compounds	Air Biota, Man
3.	NO_{3s}.No_{2s} Nitrogen oxides	Drinking water Air
4.	Mercury compounds Lead and cobalt	Food, Water Food, Air
5.	Petroleum hydrocarbons Carbon monoxide	Sea Air
6.	Fluorides	Water (fresh water)
7.	Asbestos Arsenic	Air Drinking Water
8.	Mycotoxins and microbial contaminants	Food

TYPES OF POLLUTION

Pollution may be classified on the basis of the type of environment that is being polluted. Thus we may have air pollution, water pollution and soil or land pollution. Another classification of pollution may be on the basis of type of pollutants such as Sulphur dioxide pollution, fluoride pollution, carbon monoxide pollution, smoke pollution, lead pollution, mercury pollution, solid waste pollution, radioactive pollution, noise pollution etc. Organisms do have the capacity to deal with certain amount or level of pollutants without suffering ill effects. The level of a pollutant below which no ill effects are observed is called the **threshold level**. Above the threshold level effects begin to be observed. However, the effects caused by a pollutant depends on its concentration and time of exposure. Higher levels may be tolerated if the time of exposure is less. Thus the threshold level may be high for shorter exposure, but gets lower as the time of exposure increases. There are three factors which determine the level of exposure:

(*i*) Inputs of pollutants into the air

(*ii*) Amount of space into which pollutants are dispersed and diluted

(*iii*) Mechanisms by which these are removed from the air.

Some compounds that bioaccumulate have extremely low threshold levels, and radio-active compounds have a zero threshold level, in such case any exposure, regardless of how small, may cause damage. As a part of natural ecosystem, we human beings have always relied on the different natural processes for the disposal of different waste materials we have been producing. But with the tremendous increase in human population and with equal stress on natural resources, the situation has become extremely unbalanced. The amount of the biodegradable material has become so high that the natural processes of decomposing it is not able to do so. This situation has further been aggravated by a huge amount of non-biodegradable material which our civilized industrial society is dumping on the earth every day.

SOIL POLLUTION

Soil which is formed by the weathering of rocks, is a very important resource. Various activities of human beings leading pollution of soil not only in urban and industrial areas but also in rural areas. Dumping of various waste generated in various industries and municipalities causes release of various toxic substances which ultimately find their ways in to the soil and ultimately to the ground water. The use of various types of pesticides, herbicides, insecticides, fertilizers etc. to increase the crop production also leads to the severe contamination of soil. In the soil pollution the different pollutants remain in direct contact of soil for a longer period and enter in the food chain of the nature and also in the air and water.

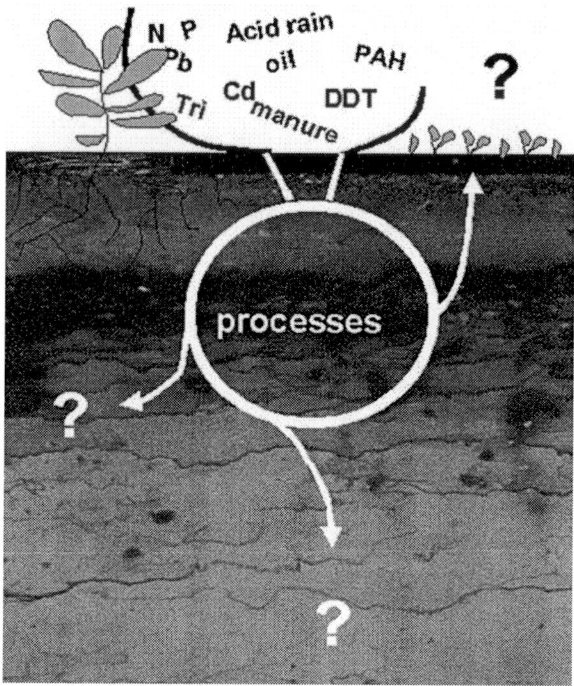

SOURCES OF SOIL POLLUTION AND THEIR EFFECTS

The important sources of soil pollution are as under.

(*i*) **Agriculture practices**: The use of large quantity of fertilizers, insecticides, herbicides, weedicides and various soil conditioning agents pollute the soil to a great extent. Similarly excessive amount of plant and animal wastes is also adding to the problem of soil pollution. The fertilizers used to improve the fertility of soil also contaminate it with their impurities. The excessive use of nitrogen, phosphorus and potassium also enhances to this problem.

The increasing use of pesticides to control various pests is causing a stress on the natural environment. During last one decade the number of pesticides has increased manifolds and important among these are organophosphates like malathion, parathion, ethion, trithion etc., and the chlorinated hydrocarbons *e.g.*, BHC, DDT, aldrin, lindane, chlordane and endosulphan. When these pesticides are applied for the contreol of a pest, these are released to the soil and their remanants contaminate to the soil. The herbicides used at the time of seedling remains active for a long time thus add to the soil contamination. Various soil conditioners and fumigants used to increase and protect the soil fertility have several toxic metals which keep on accumulating in the soil.

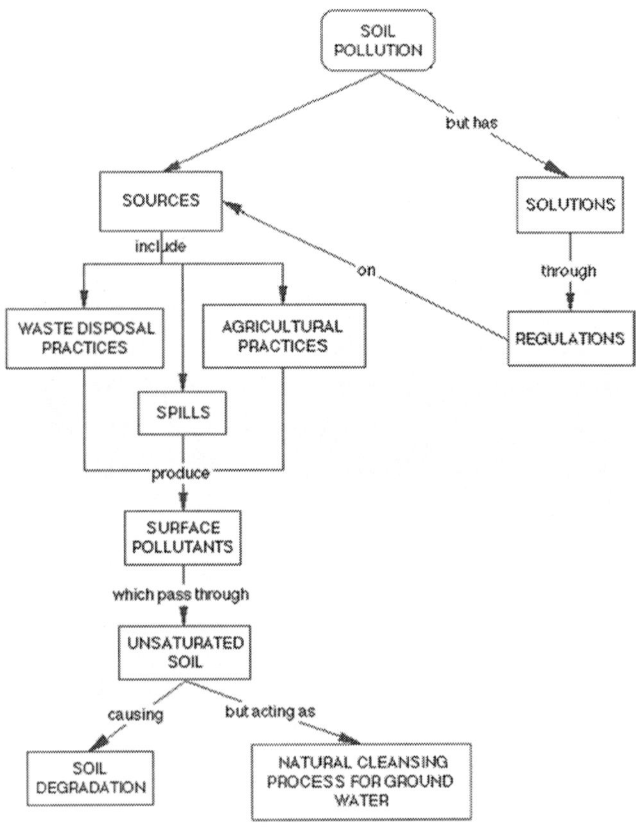

(ii) **Urban waste**: The large amount of waste being generated in the urban areas contribute to soil pollution. This waste which is normally termed as refuse contains garbage, plastics, metals, fibre, paper, glasses, containers, fuel residues etc. The generation of this waste in huge quantities has become a problem world over. The capital of India generates approximately 5000 tones of garbage everyday. The soil pollutants ultimately find their way to ground water thus affecting it severely.

(iii) **Industrial waste:** This is one of the most important soil pollutant and its disposal is becoming a serious problem with every passing day. The industrial pollutants are mainly discharged from sugar factories, oil refineries, tanneries, distilleries, fertilizer and pesticide industries and also from chemical and steel factories. The metal processing industries, drugs, cement, petroleum and coal and mineral industries are other important generators of such waste material. Fly ash, which is mainly being generated by the thermal power plants is another important soil pollutant. All the pollutants released by these industries ultimately affect and alter the biological and chemical properties of the soil.

Effects of Soil Pollutants: The various effects of soil pollutants are as under:

(i) The beneficial microorganisms such as bacteria etc. are destroyed by metallic contaminents. These contaminents such as Cr, Zn, Pb, Cd, Hg etc. remain in the soil and their accumulation in the soil for a long period is injurious to the living organisms.

(ii) Industrial wastes from fertilizers, steel, paper and textile industries are extremely toxic are sometimes very toxic to living beings. These are transferred to different organisms through the food chain and lead to a variety of harmful effects in living world.

(iii) Solid waste which is directly thrown in the open land results into offensive odour and causes a variety of diseases and ground water pollution.

(iv) Radioactive pollutants dumped into soil produce great human miseries and are responsible for a number of diseases in human beings.

(v) The various volatile materials which are released into the spoil when released into the atmosphere, contaminate it through the plants they find their way in to human beings causing disruption in the various physiological processes.

(vi) Various fertilizers and pesticides added in to the soil are responsible for many ill effects in other biota.

(*vii*) Many pathogenic bacteria which are present in soil cause serious threats to human health.

WATER POLLUTION

The sources of water on this planet are rain, surface water, the ground water and the sea. Rain water carries the washed out minerals, salts and organic matter from the earth's surface and is stored in ponds, lakes and rivers. It percolates underground and is stored there as ground water. The source of fresh water are ponds, wells, lakes and rivers. Natural water contains numerous organisms like phytoplankton, zooplanktons, fish and so on. Besides it is a good solvent and contains dissolved gases like oxygen. When we say pure water, we mean water free from organisms, particularly microbes, and which usually contains a negligible amount of salts. In India, the ever increasing population, urbanization, agricultural activities and industrialization, accompanied by a greater mechanization in every sector of life, has led to a constant increase in requirements placed upon water. It has been estimated that 70% of available water in India is polluted. The signs of water pollution have been obvious to even the most casual observer. However, water pollution may be defined as "the addition of any substance to water or changing of it's physical and chemical characteristics in any way, which interferes with its use for legitimate purposes". Thus, pollution of natural water implies that it contains a lot of inorganic substances introduced by human activities, which change its quality and are harmful for many living organisms, including man. It is one of the most serious problems in the world, particularly in developing countries. Normally water is never pure in a chemical sense. It contains impurities of various kinds-dissolved as well as suspended. These include dissolved gases (H_2S, CO_2, NH_3, N_2), dissolved minerals (Ca, Mg, Na salt), suspended matter (clay, silt, sand) and even microbes. These are natural impurities derived from atmosphere, catchments areas and the soil. They are in very low amounts and normally do not pollute water and it is potable. Polluted water, however, is turbid, unpleasant, bad smelling, unfit for drinking, bathing, and washing or other purposes. They are harmful and are vehicles of many diseases as cholera, dysentery, typhoid etc.

SOURCES, NATURE AND EFFECTS OF WATER POLLUTANTS

The chief sources of water pollution are (*i*) Oxygen demanding wastes, (*ii*) industrial effluents, (*iii*) agricultural discharges and (*iv*) industrial wastes from chemical industries, fossil fuel plants (thermal power plants) and nuclear power plants. Each of these sources of pollution carries a variety of pollutants that enter our water bodies. Following are the sources of water pollution and kind of he pollutants carried by them :

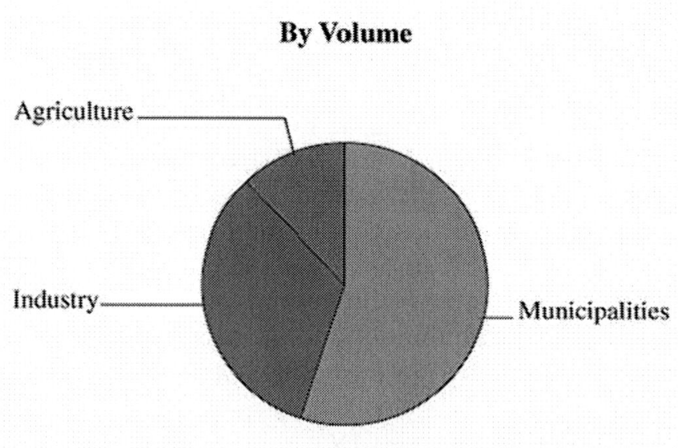

Fig. 1.4 : Major Contributors to water pollution

1. Oxygen demanding waste: Dissolved oxygen has been a fundamental requirement of life in any given body of water. A body of water is classified as polluted when the DO concentration drops below the level necessary for sustaining normal biota for that water. The primary cause of water degradation has been the presence of substances collectively called oxygen demanding wastes. These are primarily organic materials that are oxidized by bacteria to carbon di -oxide and water. Sewage is the waterborne waste derived from home (domestic waste) and animals or food processing plants. Sewage and other oxygen demanding wastes have been classified as water pollutants because their degradation leads to oxygen depletion. This affects the aquatic life, impairs domestic and livestock water supplies by affecting taste, odors and colors. It includes human excreta, paper, cloth, soap, detergents etc. There is uncontrolled dumping of wastes of rural areas, towns and cities into ponds, lakes, stream or rivers. The decomposition of these wastes by aerobic microbes decrease due to higher levels of pollution. The self-purifying ability of the water is lost and water becomes unfit for drinking and other domestic uses. Since decomposition of sewage and other wastes is largely an aerobic process, accumulation of these in water increase its oxygen requirements (BOD).

Phosphates are the major ingredients of most detergents, They favor the luxuriant growth of algae which form water blooms. This extensive algae growth also consumes most of the available oxygen from water. This decrease in level becomes detrimental to growth of other organisms, which produces a foul smell upon decay.

One of the most common primary sources of water pollution is the discharge of untreated or partly treated sewage in water bodies, sometimes due to improper sewage-handling processes of municipal bodies. As oxygen demanding wastes rapidly deplete the DO of water, it is important to estimate the amount of these pollutants in a given body of water. The biochemical oxygen demand (BOD) of water has been a quantity related to the amount of wastes present.

Biological oxygen demand (BOD): In a water sample the BOD indicates the amount of dissolved oxygen used up during the oxidation of oxygen demanding wastes. BOD is the amount of oxygen required for biological oxidation by microbes in any unit volume of water. It could be found out by incubating a sample of water for five days at 20° C. A BOD value generally approximates the amount of oxidisable organic matter, and is, therefore, used as a measure of degree of water pollution and waste level. Thus mostly BOD value is proportional to the amount of organic waste present in water.

BOD values are thus useful in evaluation of self-purification capacity of a water body and for possible control measure of pollution. The quantity of oxygen in water (Dissolved Oxygen –DO) along with BOD is indicated by the kind of organisms present in water. Thus fish become rare at DO value of 4 to 5 ppm of water. Further decrease in DO value may lead to increase in anaerobic bacteria. A BOD of 1 ppm has been characteristic of nearly pure water. Water is regarded as fairly pure with a BOD of 3ppm, and of doubtful purity when the BOD value reaches 5 ppm.

Eutrophication: In natural conditions relatively small quantities of nutrients and sediments are leached or eroded from the land into water bodies. In turn these are fed by clear rivers and streams, the natural conditions of lakes and estuaries is generally **oligotrophic,** that is the water is nutrient poor. This limits the growth of phytoplankton, but benthos plants (plants that grow attached to or rooted in the bottom) thrive to a depth of 30 feet. Oxygen from the atmosphere is very slow to dissolve and get mixed with water. Benthic plants also help in maintaining the dissolved oxygen level in deeper water because oxygen produced during photosynthesis is directly dissolved into the water. Thus a nutrient poor water body maintains a rich and diverse ecosystem.

With the erosion and leaching of materials, a water body gradually gets filled with sediments and gets enriched with nutrients. The nutrient enrichment promotes the growth of phytoplanktons which in turn make the water more turbid this problem is more compounded by sediments. The oxygen produced by photosynthesis of phytoplanktons supersaturates the upper water and escapes the surface. Phytoplankton has a high turnover rate leading to an unusually high accumulation of detritus. As decomposers mainly bacteria feed on this material, they also consume oxygen in their respiration, thus depleting the dissolved oxygen in the water. When DO is not available the bacteria shift and continue to thrive on fermentation. As a consequence, the DO becomes high at surface level while becomes almost zero at lower levels. Thus the lake becoming a eutrophic lake.

Thus eutrophication is nutrient enrichment that promotes the growth of phytoplankton in the water body.

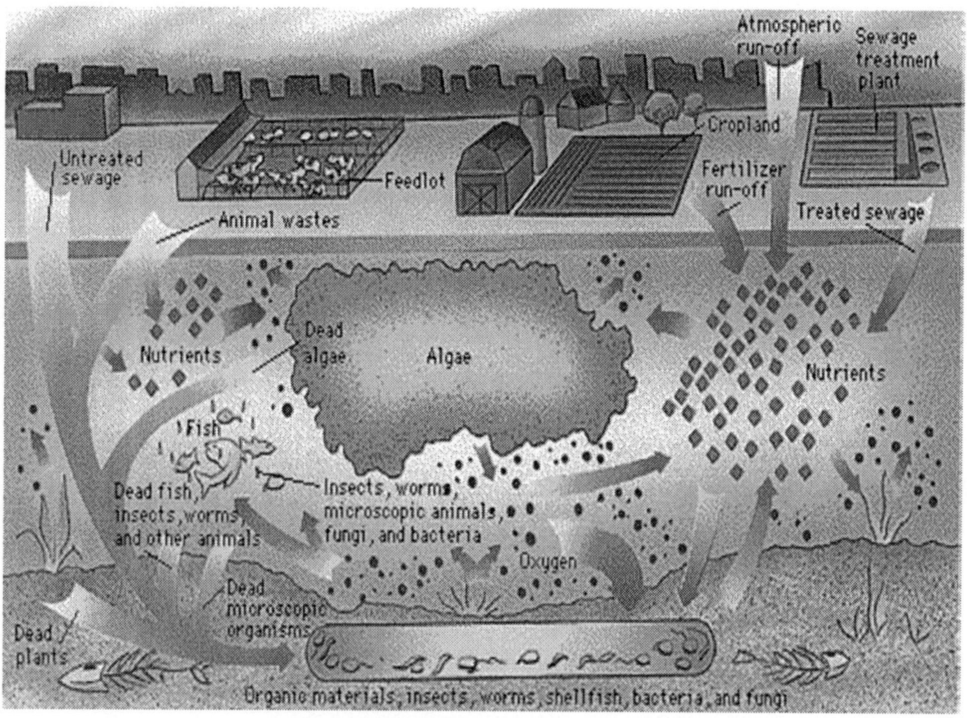

Fig. 1.5 : Major contributors to eutrophication

In USA, Lake Erie is an excellent example of eutrophication due to man made problems. In 1965, there were added more than 80 tons of phosphates daily in the lake. Each 400 g of PO_4 encourages about 350 tones of algal slime. Due to this algal growth on the shores of lake appeared as big mounds producing unpleasant odor, clogging pipes and interfering with fishing and navigation. Eutrophication is thus limiting factor in supply of clean water for drinking fishing and navigation etc.

2. Industrial wastes as water pollutants: Many industries, such as steel and paper are situated on the banks of rivers as they require huge amount of water in their manufacturing process. These industries dump their wastes in to rivers. A wide variety of both, inorganic and organic pollutant are present in effluents from breweries, tanneries, dying textiles, paper and pulp mills, steel industries, mining operations etc. The pollutants include oils, greases, plastics, metallic wastes, suspended solids, phenols, toxins, acids, salts, dyes, cyanides, DDT etc., many of which are not readily susceptible to degradation and thus cause serious pollution problems. H_2SO_4 as acid waste from coal mines is a serious pollutant that increase the hardness of water, has disastrous effect on live organisms and corrodes concrete etc. Na, Cu, Cr, Cd, Hg, Pb, etc. are the heavy metal effluents, discharged from industries. Most of the Indian rivers and fresh water streams are seriously polluted by industrial wastes and effluents which come along waste waters of different industries such as petrochemicals, fertilizer factories, oil refineries, pulp, paper textiles, sugar and steel mills, tanneries, distilleries, synthetic material plants for drugs, fibers, rubbers, plastics etc.

3. Agricultural discharges : It includes sediments, fertilizers, pesticides and farm animal wastes. All these pollutants can enter waterways as run off from agricultural lands but farm animal wastes are an especially large problem near the large feed lots on which thousands of animals concentrated. Their discharge reach into the water bodies. As compared to developed nations, India has a relatively low use of chemicals, thus discharge into water are still low. India uses about 16 Kg/ha of fertilizers (chemicals) on an average, whereas the world average is 54 Kg/ha. Thus it is not only the increased use but also escalated production that would enhance pollution.

(*a*) Artificial fertilizers: Modern agriculture relies heavily on a wide range of synthetic chemicals which include different types of fertilizers and biocides (pesticide, herbicides or weedicides). These chemicals along with wastes are wasted off lands through irrigation, rainfall, drainage etc. reaching into the rivers, lakes, stream etc. where they disturb the natural ecosystem.

How dangerous are they? Are they harmful? Do they benefit? Are they safe? What to use? Which to use? While improving the yield do these chemicals in any manner affect the quality of the food stuffs grown?

Artificial fertilizers crowd out useful minerals naturally present in the topsoil. The microbes (bacteria, fungi, worms etc.) in topsoil enrich the humus and help to produce nutrients to be taken up by the plant and later by animals. But fertilizers enriched soil can not support microbial life and hence there is less humus and less nutrients and the soil can easily become poor and eroded by wind and rain.

Chemical fertilizers are made up of only a few minerals. Thus they impede the uptake of other minerals and imbalance the whole mineral pattern of plant body. Many crops today lack potassium due to excessive use of nitrogenous fertilizer. Excessive potash treatment decreases valuable nutrients in foods, such as ascorbic acid (Vitamin C) and carton. Liming can prevent the release and uptake into the plants of cobalt, nickel, manganese and zinc. Super-phosphate may lead to a copper and zinc deficiency.

Plants also become less resistant to diseases. NO_3 fertilizers increase the total crop yield (carbohydrates) but at the expense of protein. Corn and wheat grown on soil fertilized with N, P, and K showed a 20-25% decline in protein content and increase in carbohydrate content. Moreover, subtle balance of amino acid is disturbed within the protein molecule, thus lowering protein quality. Since most of Indians are vegetarians, consumption of low quality protein leads to malnutrition. Fertilizer use produces over-sized fruits and vegetables, which are more prone to insects and other pests.

According to H. H. Koepf, an eminent soil chemist, modern agriculture can honestly claim only two notable crops-"**disease and pests**". To this we can now add a third- **poison** (as nitrites, nitrates). Nitrate fertilizers used on soil enter our wells and ponds, making their water very rich in nitrates. It not only makes water unfit for drinking but also causes disease. This water when taken by us, the nitrates are converted to nitrites by microbial flora of intestine. These nitrites then combine with the hemoglobin of blood to form methaemoglobin, which interfaces with the O_2-carrying capacity of the blood. The disease produced in called **Methaemoglobinaemia**. This leads to various ailments, damage to respiratory and vascular system, blue coloration of skin and even cancer. A healthy person contains 0.8% of methaemoglobin, whereas in methaemoglobinaemia this level reaches to 10% in the blood. At above 30% concentration there occurs headache, giddiness and above 60%, un-consciousness with stiffness and ocular problem etc takes place. At 80% death occurs. Nitrate poisoning is frequent in Rajasthan due to hard and saline water. Several children have died due to this problem. In 1976, there were cases of NO_3 poisoning of cattle in Nagpur. In Rajasthan NO_3 levels in water area very high, (800 mg/l), which is much beyond the permissible concentration of 45 mg/l by WHO. Consumption of vegetable grown in NO_3 rich soil may also lead to this disease, especially in sick persons and children. Addition of phosphates and NO_3 to water leads to depletion of oxygen due to excessive algal growth. It leads to death of fish and other aquatic life.

***(b)* Pesticides and biocides :** Pesticides are the chemicals used for killing the plant and animals pests. It is a general term that includes bactericides, fungicides, nematicides, insecticides and also the herbicides or weedicides. Since weeds (herbs) are not pests like bacteria, fungi, nematodes, insects, the spectrum of activity of these chemicals is extended beyond the pest; and thus a broader term biocide is used to include also herbicides etc.

A wide range of chemicals is used as biocides. But the most harmful are those, which either do not degrade or degrade very slowly in nature. We prefer to distinguish such chemical substances as **hazardous** substance or **toxicants.** These are highly potent chemicals that enter our food chain and then begin to increase in their concentrations at successive trophic levels in the food chain. Equally hazardous substances are the radionuclides. The hazardous biocides cause considerable harm since their effects are cumulative. Most nations have banned the use of some of these biocides.

The long range effects of such biocides are infact a threat to our ecological security. According to Pearson (1985) pesticide related deaths in developing countries are estimated at 10,000/year and about 1.5-2 million person suffer from acute pesticide poisoning.

Some of the most toxic biocides are DDT (dichloro-diphenyl trichloroethane), BHC (benzene hexachloride), chlordane, heptachlor, methoxychlor, toxaphene, aldrin, endrin an PCBs (polychlorinated biphenyl's). Indiscriminate use of the biocides could make them an integral part of our biological, geological and chemical cycles of the earth. They are everywhere in same form. Measurable amounts of DDT residues may be found in air, soil, and water and at several thousand of Kms. from the point where it had originally entered the ecosystem. For instance if DDT enters a pond, lake, it is taken as such by the plants of the pond, then reaches to zooplankton feeding on plants, then to minnows eating the zooplanktons, then to fish which eats the minnows and finally in the body of birds who eat the fish. Not only DDT as such in its original form keeps on moving form water to different living components of the pond system but more threatening is that–DDT concentration continuously increases in successive trophic levels (various form of living organisms) in a food chain. This phenomenon is known as **biological magnification** or **biological amplification**. This is the reason why our food grains as wheat, rice, vegetables and fruits today contain varying amounts of pesticide residues, which have become their integral part. They can not be removed by washing or other means.

Besides DDT there are also heavy metals like lead, mercury, copper which also show similar behavior in a food chain. Similarly the radionculides as storntium-90 follows biological magnification. We shall present more details of these processes in separate chapter on Toxicology. The outcome of such criminal and indiscriminate use may be acute (immediate) or chronic poisoning. The danger of long term consumption of pesticides residues in food is far more serious that acute poisoning from the point of view of national health. Children born today have to start life with a body burden of pesticides, which increase with age. There is evidence that such chronic accumulation of pesticides played a role in kidney malfunctioning, excess of amino acids in blood and urine, electroencephalogram abnormalities of brain tissue, blood abnormalities etc.

4. Solid waste pollution: Solid waste varies in composition with socioeconomic status of the generating community. The disposal of solid waste poses many problems, depending upon both the types of waste and the disposal method employed. The majority of waste classified as combustible- rubbish, garbage and sewage sludge is disposed off by incineration, using it as land fill or disposal by ocean dumping. Landfill and ocean dumping lead to water pollution.

5. Thermal pollution: Thermal pollution takes place because many electricity generating companies use water in the process of cooling their generators. This heated water is then released into the system from which it was drawn, causing a warming trend of the surface waters. Thermal pollution results when the heated effluent is released into poorly flushed systems. In these cases a permanent temperature increase often results, which tends to decrease the solubility of dissolved oxygen in the waterbody and is thus very injurious to aquatic life.

6. Radioactive waste pollution: The major sources of radioactive wastes have been nuclear explosives, accidents at nuclear power plants, fuel reprocessing plants and research laboratories and hospitals that release these wastes into the atmosphere or into waste water.

Ground Water Pollution

Out of total fresh water present on the earth, 98% lies below its surface. The remaining 2% is available in the form of rivers, streams, ponds and lakes. Ground water occurs in saturated materials below the water table. It acts as a reservoir by virtue of large pore space in earth materials, as a conduit which can transport water over long distances and as a mechanical filter which improves water quality by removing suspended solids and bacterial contaminants.

Because of the various anthropogenic activities many industrial, domestic and agricultural wastes are being constantly added in the ground water and the rate of its pollution is very high. Ground water receives pollutants frm the following important sources:

(*a*) Domestic waste: This includes pathogenic organisms, nutrients and solids and also the oxygen demanding materials.

(*b*) Industrial wastes: These include toxic heavy metals along with organic and inorganic hazardous materials. The industries of woolens, bicycles in areas of Punjab contribute high amount of Ni, Fe, Cu, Cr and cyanides to ground waters.

(*c*) Agricultural wastes: Many fertilizers, pesticides and insecticides added to the soil to improve its fertility and protect the crops from pests and rodents find their ultimate way in to the ground water thus pollute it. Similarly processing waste and animal wastes are also added to the water. Leachets from agricultural land containing nitrates, phosphates and potash move downward with percolating water and ultimately get mixed with ground water thus polluting it.

(*d*) Seepage pits, Urban and rural garbage, waste water treatment lagoons are some pther potential sources of underground water pollution.

Effects of Ground Water Pollution

Ground water pollution not noly affects the human beings but also to various other objects and materials on which we are dependent. The important effects are as under:

(*a*) It is a major cause of spreading epidemics and chronic diseases like typhoid, cholera, dysentery and diarrhoea etc. in human beings.

(*b*) Contaminents like asbestos are responsible for causing fatal diseases like asbestosis and lung cancer.

(*c*) Ground water in excessive rainfall areas contains iron toxic amounts.

(*d*) Cropsirregated with polluted ground water get severely damaged and productivity of grains also decreases.

(*e*) Soil fertility is also affected by ground water pollution.

(*f*) Plant metabolism is also affected.

Marine pollution: Discharge of waste substances into the sea resulting in harmful effects to living resources, hazardous to human health, hinderance to fishery and impairment of quality for use of sea water is called as marine pollution. It is associated with the change in physical, chemical and biological conditions of the sea water. All that what is carried by rivers ultimately ends up in the seas. On their way to sea, rivers receive huge amounts of sewage, garbage, agricultural discharge, biocides, including heavy metals. These all are added to sea. Besides these discharges of oils and petroleum products and dumping of radionuclieds waste into sea also cause marine pollution. Huge quantity of plastic is being added to sea and oceans. Over 50 million lb plastic packing material is being dumped in sea of commercial fleets, whereas over 300 million lb entering through inland waterways in USA. Many marine bird ingest plastic that causes gastro-intestinal egg shell and tissue damage of egg. Radionuclide waste in sea includes Sr-90, Cs-137, Pu-239, Pu-240.

The pollutants in seas may become dispersed by turbulence and ocean currents or concentrated in the food chain. They may sediment at the bottom by processes like adsorption, precipitation and accumulation. Bioaccumulation in food chain may result into loss of species diversity. The pollution in Baltic sea along the coast of Finland, took place largely from sewage and effluents form bottom fauna. There was seen distinct zonation with extent of pollution. In clear or less polluted water there was rich species diversity which tended to decrease with increasing pollution load. In heavily polluted area, macroscopic benthic animals were absent, but chironomid larvae occurred at the bottom.

In marine water the serious pollutant is oil, particularly when afloat on sea. An spill in oil or petroleum product due to accidents or a deliberate discharge of oil polluted waste brings about pollution. About 285 million gallons of oil are spilled each year into ocean, mostly from transport tankers. This is enough to coat a beach 20 feet wide with half and inch oil layer for 8633 miles. About 50,000 to 250,000 birds are killed every year by oil. The oil is soaked in feathers, displacing the air and thus interfere with buoyancy and maintenance of body temperature, Hydrocarbons and benzpyrene accumulate in food chain and consumption of fish by man may cause cancer. Detergents used to clean up the spill are also harmful to marine life.

Mercury pollution: Mercury enters water naturally as well as through industrial effluents. It is a potent hazardous substance. Both, inorganic and organic forms are highly poisonous. Methyl mercury gives off vapor. Mercury was responsible for the minimata epidemic that caused several deaths, in Japan and Sweden. The tragedy had occurred due to consumption of heavily mercury-contaminated fish (27 to 102 ppm, average 50 ppm) by the villagers. The source of mercury to the bay was a single chloride producing plant, using $Hgcl_2$ as a catalyst. In Sweden many rivers and lakes are already polluted due to widespread use of mercury compounds as fungicides and algaecides in paper and pulp industries and in agriculture. Chloral alkali plants seem to be the chief source of mercury containing effluents. Paper and pulp industries of Japan and Canada also cause mercury pollution. Effluents of industries making switches, batteries, thermometers, fluorescent light tubes and high intensity street lamps also contain mercury.

Mercury compounds enter the water body from the effluents and at their bottom these are metabolically converted into methyl mercury compounds by anaerobic microbes. Methyl mercury is highly persistent and thus accumulates in food chain. Methyl mercury is soluble in lipids and thus after being taken by animals it accumulates in fatty tissues. Fish may accumulate the methyl mercury ions directly. There may be nearly 3000 times more mercury in fish than in water. In Minamata bay all the mercury in sea food is as organic methyl mercury compounds. The symptoms of Minamata include malaise, numbness, visual disturbance, dyspepsia, ataxia, mental deterioration, convulsions and final death. Mercury readily penetrated the central nervous system of children born in Minamata causing teratogenic effects. Methyl mercury penetrates through placenta. Swedish fish eaters have high mercury content in blood. In *Drosopial* methyl mercury (0.25 ppm) treatment brought chromosomal disjunction in gametes.

Mercury poisoning is caused due to inactivation of several sulfhydral enzymes by replacement of hydrogen atoms in sulfhydral groups. The antidote, BAL (dimercaprol) is used for mercury poisoning.

Lead pollution: Lead poisoning is common in adults. The chief sources of lead to water are the effluents of lead and lead processing industries. Lead toys may be chewed by children. Painters also have a risk of lead consumption. In some plastic pipes lead is used as stabilizer. The water may become contaminated in these pipes. Lead is also used in insecticides, food, beverage, ointments and medicinal concoctions for flavoring and sweetening.

Lead pollution causes damage to liver and kidney, reduction in hemoglobin formation, mental retardation and abnormalities in fertility and pregnancy. Chronic lead poisoning may cause three general disease syndromes (*i*) gastrointestinal disorder (*ii*) neuromuscular effects (leadlapsy)- weakness, fatigue muscular atrophy, and (*iii*) central nervous system effects of CNS Syndrome –that may result to coma and death. Lead poisoning also causes constipation, abdominal pain etc.

Fluoride pollution: Fluoride is also regularly present in water and soil besides air. The crop plants grown in high-fluoride soils in agricultural, non-industrial areas had a fluoride content as high as 300 ppm. In Haryana and Punjab, consumption of fluoride-rich water from wells caused endemic fluorisis. In Andhra Pradesh also high fluroride content water caused dental fluorisis. On an average, about 20-25 million Indian are affected with fluorisis. In our country this problem has become more severe in Rajasthan. This has already crippled about 3.5 lakh person in state. Fluorisis is prevalent in districts of Jodhpur, Bhilwara, Jaipur, Bikaner, Udazipur, Nagpur, Barmer and Ajmer. Many people in Rajasth have humped back due to high fluoride content in water sources and in arid and grains semi arid zones. In arid and semi-arid soils also fluroide content is very high. Food grains obtained from these soils are also rich in fluorides. Prolonged intake of fluoride containing water stiffens the bone joints, particularly of spinal cord. Fluoride is not absorbed in the blood stream. It has an affinity with calcium and thus gets accumulated in bones, resulting in the mottling of teeth, pain in the bones and joint and outward bending of legs from the knees-knock knee syndrome. Fluoride levels more than 0.5 ppm over a period of 5-10 years results in fluorisis terminating in crippling or paralysis. In water of most villages of Rajasthan fluoride level is higher than permissible limit of 1 mg/liter of water.

Cattle grazing around fluoride sources, as ceramic rocks, phosphate fertilizer plants and aluminum factories often develop fluorisis. The toxic effects are staining, mottling and abrasion of teeth, high fluoride levels in bone and urine, decreased milk production, and lameness. Moreover, the animals also become lethargic.

NOISE POLLUTION

Noise pollution is a direct result of technological development. Noise, with its ever increasing effects on human beings and on the environment, is defined as an acoustic fact which is unpleasant and arouses disturbing feeling or as the totality of unwanted, undesired sounds. The human ear is constantly being assailed by man made sound from all sides, and there remain few places in population areas where relative quiet prevails. What do airplanes, trains, cars, and numatic drills, and radio and television sets have in common? They all produce noise, the most dangerous pollutant of man's environment. Noise has become a permanent part of our lives these days because of the development of machinery, industry and technology. Noise harms the body and mind. Noise not only cause irritation or annoyance but it constricts your arteries, increases the flow of adrenaline and forces your heart to work faster.

The word **noise** (Latin nausea) is usually defined as unwanted or unpleasant sound that causes discomfort. Noise is also defined as "wrong sound, in the wrong place at the wrong time". Noise pollution means, "the unwanted sound dumped into the atmosphere leading to health hazards".

Even if noise is not sufficiently loud to constitute direct threat to human health, continuous exposure to noise shortens human beings sleeping hours and reduces his productivity. The decrease in productivity affects both city dwellers and people who work in factories or in the industrial areas. In China, till the third century BC, noise was used as a method of torture instead of hanging men for dangerous crime. The importance of noise as a pollutant having a deleterious effect on peace of mind and beauty of environment is increasing every day.

Formerly noise was limited only to the industry. This too was not much as there were only few industries. These days there has been rapid industrial growth. Moreover, there has been population explosion, due to which there is heavy traffic, urban crowd and electric equipment (luxury items and entertainment). All these have added to the noise nuisance in environment. In our country, beside these the two other factors are the religious and social functions, which increase the gravity of situation.

Properties: There are two basic properties of sound, (*i*) loudness and (*ii*) frequency. Loudness is the strength of sensation of sound perceived by the individuals. It is measured in terms of decibels. Just sound is about 10 dB, a whisper about 20 dB, library place, 30 dB, normal conversation 35-60 dB, heavy street traffic 60-80 dB, boiler factories 120 dB, jet planes (take off) about 150 dB, rocker engine, about 180 dB. The loudest sound a person can stand without much discomfort is about 80 dB. Sounds beyond 80 dB can be safely regarded as pollutant as its harms hearing system. The WHO has fixed 45 dB as the safe noise level for a city. For international standards a noise level upto 65 dB is considered tolerable. Intensity of some noise sources is as follows :

Sources	Intensity (dB)
Breathing	10
Broadcasting studio	20
Soft whisper	20-30
Trickling clock	30
Library	30-35
Low volume radio	35-40
Normal conversation	35-60
Telephone	60
Office noise	60-80
Alarm clock	70-80
Traffic	50-90
Trunk	90
Motor cycle	105
Lion's roar (12')	105-110
Jet fly (over 1000')	100-110
Train whistle (50')	110
Air craft (100') (Prokeller driven)	110-120
Pneumatic drill	110-120
Commercial jet (air craft (100'))	120-140
Jet take off (300')	120
Space rocket (launching)	170-180

Loudness is also expressed in **sones.** One sone equals the loudness of 40 dB sound pressure at 1000 hz.

Frequency is defined as the number of vibrations per second. It is denoted as **Hertz** (hz). One hz equals to one vibration per second. People can hear sound from 16 (infra-audible) to 20,000 (ultrasonic) hz.

Sources and Effects of Noise

Sources of noise are numerous but these may be broadly classified in to two classes: (*i*) Industrial and (*ii*) Non-industrial.

The main contributors to noise are factories and industries, transportation (air, rail and road). The disturbing qualities of noise emitted by industrial premises are generally its loudness, its distinguishing feature such as tonal or impulsive components and its intermitancy and duration.

Among the non-industrial sources, important ones are as follows (*a*) Loudspeakers, (*b*) automobiles, (*c*) aircrafts (*d*) trains (*e*) construction works (*f*) radio, microphone etc. The chief man-made sources in urban areas are automobiles, factories, industries, trains, airplanes. Noise makers are horns, sirens, lawn mover, musical instruments, TV, radio, transistor, telephone, dog, loudspeaker, washing machines, vacuum cleaner, food mixers, pressure cookers, fans, air conditioners, coolers. Ever since the industrial revolution, there has been doubling of environmental noise in every 10 years.

Effects: Noise has been found to interfere with our activities at three levels : (*i*) audiological level (*ii*) biological level, interfering with the biological functioning of the body and (*iii*) behavioral level, affecting the sociological behavior of the subjects.

1. Auditory effects : These include auditory fatigue, and deafness. Auditory fatigue appearing the 90 dB and may be associated with side effects as whistling and buzzing in ears. Deafness can be caused due to continuous noise exposure. Temporary deafness occurs at 4000-6000 hz. Permanent loss of hearing occurs at 100 dB.

2. Non-auditory effects : These are (*i*) interference with speech communication, (*ii*) annoyance, (*iii*) loss of working and (*iv*) physiological disorders

(*a*) **Interference with speech communication :** A noise of 50-60 dB commonly interferes with speech; sound of warning (signal) may be misunderstood.

(*b*) **Annoyance :** Balanced person express great annoyance at even low level of noise as crowd, highway, radio etc. The effects are ill temper, bricking.

(*c*) **Loss in working efficiency :** There develop tiredness and those doing mental work may put to deterioration in their efficiency or even complete loss of ability to work.

(*d*) **Physiological disorders :** A number of physiological disorders develop due to imbalance in functioning of the body. These are neurosis, anxiety, insomnia, hypertension, hepatic diseases, behavioural and emotional stress, increase in sweating, giddiness, nausea fatigue etc. Noise also cause visual disturbance, and reduces depth and quality of sleep thus affecting overall mental and physical health. Other effects are, undesirable changes in respiration, circulation of blood in skin and gastrointestinal activity. Noise pollution also causes incidence of peptic ulcers.

Continuous noise causes an increase in cholesterol level resulting in the constriction of blood vessels making you prone to heart attack and strokes. There may be still births and usually low weight children born to mothers living near airports.

Supersonic air planes create a shock wave called **sonic boon**, which produces a startle effect that can be more harmful than a continuous noise. The sonic boon may spread in an area of 10 to 80 miles and when its hits the ground it damages window pans and building structures. This may also fasten the human fetus heart rate. Some of the important health hazards of noise are as follows :

Noise intensity (dB)	Health hazards
80	Annoyance
90	Hearing damage
95	Very annoying
110	Stimulation of reception in skin
120	Pain threshold
130-135	Nausea, vomiting, dizziness
140	Pain in ear
150	Burning of skin
160	Rupture of tympanic membrane
180	Major permanent damage in short time

Control of Noise Pollution

All noise control problems could fundamentally be denoted as shown in following figure

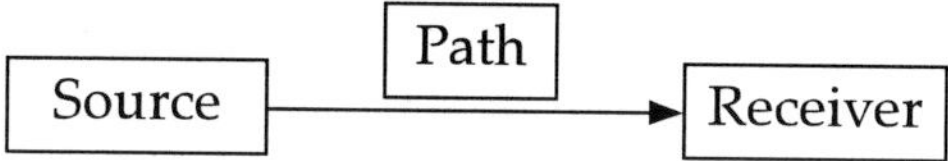

A source or sources produce the noise, and both, direct and indirect paths exist by which the generated noise could be transmitted to one or more receiver. There are following ways to control and reduce the noise menace.

1. Control at source: This can be done by (*i*) designing and fabricating silencing devices in air-craft engines, automobiles industrial machines and home appliance and (*ii*) segregating the noisy machines. The proper design of equipment to minimize the noise generation has been somewhat a complex engineering problem needing a strong background in the fundamentals of vibration, fluid mechanics, dynamics, materials and machine design. It is also important to operate all the equipment at the design conditions. Operating equipment at designed pressure and speed should result in minimum noise generation. It should be apparent that maintenance is an important element in controlling the noise generated by equipment.

2. Transmission control : This can be achieved by covering the room walls with sound absorber as acoustic tiles and construction of enclosures around industrial machinery.

3. To protect exposed person : The worker exposed to noise can be provided with wearing devices as ear plugs and ear muffs.

4. To create vegetation cover : Plants absorb and dissipate sound energy and thus act as buffer zone. Trees should be planted along high ways, streets and other places. Ashok, Neem, Tamarind etc. are good for this purpose.

5. Control through law: Silence zones must be created near schools, hospitals and indiscriminate use of loudspeakers at public places must be prevented by laws. Adequate restriction must be put on unnecessary use of horns and vehicles plying with out silencers. There are already such laws in some countries as UK and USA. In India, we have Motors vehicles Act which provides restriction on trucks using double sirens while passing through some localities. But this is not enough. In Delhi and Bombay, there are flights round the clock at airport. Delhi is closely following Bombay in noise pollution and if adequate steps are not taken to reduce sound level, more than 50% of Delhi will be affected in coming years. Restriction may be put on air craft flight at mid-night.

6. Education : No doubt we have the technology to control nearly every kind of noise. We know very well about the sources and impact of noise pollution. But the main problem is that the people are not aware that the noise is one of the main environmental problems. Public must be made aware and educated about noise nuisance through adequate news media, lectures and other programs. The movement against noise pollution is very weak in India. The main reasons being that most of us do not consider noise as a pollution but as a part of routine life.

PREVENTION AND CONTROL OF POLLUTION

In our country, the Ministry of Environment and Forests adopted a policy statement for abatement of pollution in 1992, which interalia provides several instruments in the form of regulations, legislation, agreements, fiscal incentives and various other measures to prevent and abate pollution of air, water, noise and land. After the adoption of the Policy Statement, the Ministry and its associated offices have focused various programs and schemes to prevent and control of pollution at the beginning of the pipeline by adoption of cleaner technologies, waste minimization and resource preservation rather than the traditional treatment at the end of the pipe line. The significant benefits in this approach is that when waste is reduced or solvents are revived, it leads to resources conservation of the raw materials used during the various industrial processes and minimize the pollution in the ambient environment.

In our country there have been several legislative measures, even since pre-independence period, both at State and Central Government levels to prevent and control different components of the environment. For instance, the Indian Fisheries Act, 1897, the Indian Ports Act, 1907, Bengal Smoke Nuisance Act, 1905. The Motor Vehicles Act, 1983. The Maharashtra Prevention of Water Pollution Act, 1953; The Orissa River Pollution and Prevention Act, 1954; The Prevention of Food Adulteration Act, 1985; The River Boards Act, 1956; The Atomic Energy Act (Radiation Protection Rules), 1962; The Gujarat Smoke Nuisance Act, 1963 and several other Acts promulgated from time to time.

However, the real awareness about environmental protection at global level was recognized at the United Nations Conference on the Human Environment held at Stockholm (Sweden) in June 1972. This conference was able to focus the attention of international community of environmentalists to tackle environmental problems efficiently. In this conference, our late Prime Minister Mrs. Indira Gandhi took keen interest and initiative to take appropriate steps for the protection and improvement of human environment.

Keeping in view the need for environmental protection, the Indian Parliament passed the Water (Prevention and Control of Pollution) Act, 1974, which became effective from 23 March 1974. The Central Board for Prevention and Control of Water Pollution (CBPCWP) was constituted in Sept. 1974. The Air (Prevention and Control of Pollution) Act, 1981 was promulgated on 16 May, 1981. The Environment (Protection) Act, 1986 was promulgated on 23 May, 1986.

There had already been State Boards to advise the respective State governments on matters related to water and air pollution. The CBPCWP coordinats the activities of the State Pollution Control Boards. While amending the Water (Prevention and Control of Pollution) Act, 1974 in 1988, the Central/ State Boards for the Prevent and Control of Pollution were renamed as Central/ State Pollution Control Boards as these Boards also deal with air pollution. Consequent to this amendment, there is now a Central Pollution Control Board (CPCB) that makes recommendations and advises the Central Government on pollution matters. The Board advise the Pollution Control Division of the Ministry of Environment and Forests, New Delhi. There are also State Pollution Control Boards (SPCB) to advise respective State governments.

(*i*) Enforcement of Act: The best protection of the environment is not to generate pollutants in the first place, non polluting manufacturing processes can not be added on to existing units but must be planed from the outset. The Air (Prevention and Control) Act of 1981 was enacted by invoking the Central

Governments power under Article 253 to make laws implementing decisions taken at international conferences. The preamble of the Air Act states that the act represents an implementation of the decisions made at United Nations Conference on the Human Environment at Stockholm in 1972. The Air Act has been amended in 1987. The Act enables a magistrate to restrain an air polluter from discharging emissions, and empowers both the central and State boards to give directions to industries, which if not followed, can be enforced by the board closing down the industry or withdrawing its supply of power and water. The penalties have been increased so that the polluter's cost of non compliance is substantial. Finally, citizens can not only sue to enforce the act to gain compliance by the industry, but can also require the board to provide the data needed to build a citizen's case.

The one initiative in formulating national pollution control standards began about 1984, when the central department of environment began to promulgate minimum national standards (MINAS). MINAS have been promulgated both for the water and for air.

(*a*) The Water (Prevention and Control of Pollution) Act, 1974 represented one of India's first attempt to deal comprehensively with an environmental issue. Parliament adopted minor amendments to the Act in 1978 and revised the Act in 1988 to more closely conform the provisions of the Environment (Protection) Act 1986. The water Act is comprehensive in its coverage, applying to streams, inland waters, subterranean waters, and sea or tidal waters. Standards for the discharge of effluent or the quality of the receiving waters are not specified in the Act itself. Instead, the Act enables the state boards to prescribe these standards. The state boards must maintain and make public a register containing the particulars of the control orders. The Act empowers a state board, upon thirty days notice to a polluter, to execute any work required under a consent order which has not been executed. The board may recover the expenses for such work from the polluter.

(*b*) The Environment (Protection) Act, 1986: The Act was promulgated to provide for the protection and improvement of environment and matters connected therewith. The Act consists of 26 sections distributed among four chapter, and extends to the whole of India. Section 2 defines the various terms as follows :

"environment" includes water, air and land the interrelationship which exists among and between water, air and land, and human beings, other living creatures, plants, microorganisms and property.

"environmental pollutant" means any solid, liquid or gaseous substance present in such concentration as may be, or tend to be, injurious to environment;

"environmental pollutant" means the presence in the environment of an environmental pollutant;

"hazardous substance" means any substance or preparation which, by reason of its chemical or physio-chemical properties or handling, is liable to cause harm to human beings, other living creature, plants, micro-organisms, property or the environment.

The Act provides general powers to the Central government to take all necessary measures for the purpose of (*i*) protecting and improving the quality of the environment, and (*ii*) preventing, controlling and abating environmental pollution. Besides other powers, the Central government shall have power for (*i*) planning and execution of a nation-wide programme for the prevention, control and abatement of environmental pollution, (*ii*) laying down standards for the quality of environment in its various aspects, (iii) laying down standards for emission or discharge of environmental pollutants from various sources whatsoever, (*iv*) restriction of areas in which industry, operations or processes shall not be carried out or shall be carried out subject to certain safeguards, (*v*) laying down procedures and safeguards for prevention of accidents which may cause environmental pollution, (*vi*) laying down procedure for handling of hazardous substances, (*vii*) examination of such manufacturing processes, materials and substances as are likely to cause environmental pollution, (*viii*) carrying out and sponsoring investigations and research relating to problems of environmental pollution, (*ix*) collection and dissemination of information on environmental pollution and (*x*) preparation of manuals, codes or guides relating to the prevention, and abatement of environmental pollution.

The Air (Prevention and Control of Pollution) Act, 1981 was amended in 1987 to remove the difficulties encountered during implementation, to confer more powers on the implementing agencies and to impose more stringent penalties of the provisions of the Act. The main point was also to amend the definition of air pollutants to include noise also. There is thus also the Air (Pollution and Control of Pollution) Amendment Act, 1987.

The Water (Prevention and Control of Pollution) Act, 1974 was also amended in 1988. An important amendment was to rename the Central/State Boards for Prevention and Control of Water Pollution as Central/State Pollution Control Boards as Boards also deal with air pollution. Some more powers were given to CPCB. The Boards have been given power to close or stop supply of water and electricity to offending establishments. The citizens may file criminal complaints against offenders after 60 days notice to Boards. Even at the time of establishment of industry the person will have to take consent of the Board. There is thus also the Water (Prevention and Control of Pollution) Amendment Act, 1988.

(*c*) The motor vehicles act, 1988: This Act has come into forces from Ist July 1989. The salient features of the provisions of the Act, besides other things also include the requirements to be observed in matter of vehicles fitness. Some of the features are as follows :

1. For registration of a new vehicle, the vehicle is to be produced before the registering authority for inspection to satisfy that the vehicle is fit for registration. The certificate of fitness is to be issued for a new vehicle for a period of two years. Thereafter, the renewal shall be for a period of one year till the vehicle attains the age of 10 years.
2. Regarding construction, equipment and maintenance of motor vehicles, from 1.10.1989 every motor vehicle shall comply with the emission standards.

Method of Test	Maximum smoke density		
	Light ml	Bosch unit	Hartidge
Absorption Coefficient			
(i) Full load at a speed of 60% to 70% of maximum engine rate speed declared by the manufacturer	3.1	5.2	75
(ii) Free acceleration	2.3	—	65

3. Transportation of goods of dangerous or hazardous nature to human life has been regulated under the Act. The vehicle carrying such goods should have prescribed labels indicating the nature of hazardous goods. The packages should also bear such labels. The driver should know the fundamentals in case of the vehicle is involved in accident. Driver should have minimum X standard qualification.
4. Every vehicle is required to meet the safety standards of components. Every vehicles manufacturer shall have to certify that every part used in the vehicle complies with the standards reliable to such components laid down by Bureau of Indian Standards.
5. A prototype of the vehicle to the manufactured by the manufacturer will have to be tested by VRDE, Ahmadnagar, ARAI, Pune or CMTTI, Budni or any other agency specified by the Central Government.
6. The horn to be used is to be in accordance with the approved specifications of Bureau of Indian Standards.
7. New items are included for offences and penalties in the Act. Amounts for different types of offences have been specified.

To help the owners of car, scooter maintain the prescribed level of exhaust emissions, the Directorate of Transport in Delhi has established Free Pollution Checking Centers at each of its offices in Delhi. Besides this, mobile teams are also covering different parts of the city. Delhi Administration has also been providing free facilities at different Petrol Stations all over the city since November 1987 for checking exhaust carbon monoxide levels from cars and scooters.

***(ii)* To check vehicular pollution:** Motor vehicles are one of the most important sources of air pollution. Pollution of air environment can be minimised by controlling the emissions from the vehicles. This can be achieved by (*i*) using proper proportion proportion of gasoline and air, (*ii*) more exact timing of fuel feeding, (*iii*) using gas additives to improve combustion, (*iv*) by injecting air into the exhaust to convert exhaust compounds to less toxic materials, and by (*v*) updating of engine design and/or install abatement equipment to improve combustion with the existing engine design. Carbon monoxide results from low air content of the fuel mixture, whereas NO_x production is promoted by high combustion temperatures. A hydrocarbon follows more or less the pattern of CO. The complete elimination of these three pollutants may be achieved by either updating the present design of engines or by making appropriate changes in devices for improving combustion.

Some gas vapours escape between walls and the piston which enter the crankcase and the discharge into atmosphere. Hydrocarbons (about 25%) are released in this way. Thus use of filters that capture and recycle these escaped gases in the engine should help control the emission of these hydrocarbons.

(*iii*) Lead free petrol: Lead free petrol must be available at a cheaper rate so that vehicle users can be motivated towards toxic free fuel. '

(*iv*) Industrial pollution:To check air pollution by industrial and power plant chimney wastes, we must device measures for removal of the particulate mater and gaseous pollutants from the wastes. Removal of particulate mater involves their collection under the influence of different forces, thereby moving them continuously out of the gas stream. The equipment used for their removal are: (*i*) cyclone collector, and (*ii*) electrostatic precipitators (ESPs). Thus we have to generate the control technology. At present there are few power plants and industries that have installed the requisite ESPs. In cyclone collectors the waste gas containing particles is subjected to centrifugation. The suspended particles move towards the wall of cyclone body, and then to its bottom and finally discharged out. The cyclone collectors remove about 70% of the particles. Electrostatic precipitators (ESPs) are used to remove the particles from gas stream. Suspended particles become charged or ionized, and they are attracted to charged electrodes and removed. ESPs can remove 99% of the particulate pollutants from chimney exhaust. ESPs work very well in power plants, paper mills, cement mills, carbon block plants etc.

(*v*) Alternative sources of energy: The alternative sources of energy such as wind, solar and water may be taken into account in the consideration of various solutions to air pollution problems.

(*vii*) Plantation of trees: Plantation of trees, specially , broad leafed plants help in retaining most of the gases and dust in their leaves, twigs and stems.

(*viii*) Major activities

The other major activities to control through include:

(*g*) Taj Protection Mission: Consequence upon the Supreme Court orders dated 4-9-96 in the writ petition no. 13381/84, the Planning Commission estimated an amount of ₹ 600 crores on a 50:50 percent (Center-State) matching basis to implement various schemes related to protection of the Taj Mahal. A Mission Management Board under the chairmanship of the Chief Secretary, UP has been constituted for overseeing the implementation, monitoring and reviewing of various programs/ schemes formulated for the protection of Taj Mahal.

(*b*) Air quality monitoring: There is no scientific and proper monitoring of various environmental pollutants in the country. This could be undertaken only very recently and information could become available for some areas. A National network of ambient air quality monitoring stations was initiated in 1984. Till to date (1989) under this network a total of 140 stations have been sanctioned. Of these 103 stations are operational covering about 24 cities/towns. Ten more station, 5 each for Bombay and Calcutta are likely to be sanctioned soon. Three automatic ambient air quality monitoring stations (2 stationary, 1 mobile) were established at Delhi in 1988 under the Indo-EEC bilateral program. The parameters to be measured are SO_2, CO, NO_x, SPM, temperature, humidity, wind speed and directions.

(*c*) Assessment of water quality : Under the National water quality monitoring programme, the water quality monitoring of rivers being done. Till 1987 this network had 200 monitoring stations on rivers all over the country. In 1988, there were added 106 station to the network. Thus at present there are 306 monitoring stations and 27 station on the Ganga under the Ganga Action Plain.

(*d*) Assessment of coastal water quality : The CPCB in collaboration with the Dept. of Ocean Development has identified 173 monitoring station all along the Indian Coast to assess to water quality. Four SPCB have also been involved. Data on 25 parameters are being processed to schemes to control and monitor pollution of the coastal waters.

(*e*) Preparation of environmental standards : Based on the standards prepared by the CPCB and the Bureau of Indian Standards (BIS), effluent and emission standards for different kinds of industries including thermal power plants have been notified under the Environment (Protection) Act, 1986. The industries covered are, caustic soda, man made fiber, oil refinery, sugar, thermal power plants, cotton, textiles, composite woolen mills, dyes, electroplating, cement, stone crushing, coke ovens natural rubber, synthetic rubber, small pulp and paper, fermentation, tanneries, fertilizers, aluminum, copper, lead and smelting, calcium carbide, carbon black, nitric acid, sulphuric acid, iron and steel. The work on prescription of standards for emissions from vehicular pollution in 12 metro cities is in progress.

(*f*) Enforcement of standards : This is very helpful to control pollution at source. Minimum National Standards (MINAS) and air emission standards are being evolved by the CPCB for major categories of water and air polluting industries respectively. These standards refer to the maximum limit of effluents and emission that an industry may discharge into any water body or the atmosphere. The SPCBs can stipulate the same or more stringent standards for effluents and emission discharges.

National and Zonal Task Forces have been set up to implement the standards in fertilizer industries, iron and steel, cement, pulp and paper (small) and thermal power. The progress in set up of effluent treatment plants (ETPs) in different industries is also evaluated for preparation of standards along with MINAS and air emission standards. Effluent standards, prescribed by pollution control board is given as under:

Parameters	Tolerance Limit
1. pH	5.52 to 9.0
2. Suspended Solids (mg/l.)	100.00
3. Temperature max. (°C)	40 °C
4. Oil and Grease (mg/l)	10
5. Ammonical nitrogen as (mg/l) N	50
6. BOD at 20 °C(mg/l)	30
7. COD (mg/l)	250
8. Arsenic (mg/l)	0.2
9. Mercury (mg/l)	0.01
10. Lead (mg/l)	0.1
11. Cadmium (mg/l)	2.0
12. Flexavalent chromium (mg/l)	0.1
13. Copper (mg/l)	3.0
14. Nickel (mg/l)	3.0
15. Selinium (mg/l)	0.05
16. Zink (mg/l)	5.0
17. Dissolved phosphate (mg/l)	5.0
18. Dissolved fluoride (mg/l)	10.0

Emission standards for vehicular exhausts as prescribed by WHO is given as under:

Pollutants	Light Vehicle	Heavy duty Petrol vehicles	Heavy duty Diesel vehicles	Motorcycle
Particulate matter	.33	.045	.75	.2
SO_2	.68	.39	1.5	.02
NO_x	3.2	0.99	21.0	0.07
CO	40.0	.21	2.1	10.0
HC	6.0	11.0	12.7	17.0

(*g*) **Ganga action plan :** The Ganga is regarded as the cradle of Indian civilization. It has been sinking and stinking by the tones of dangerous chemical and organic wastes emptied into it daily. The govt. of India announced an ambitious new plan in 1985 for cleaning up the Ganga river. There is a separate Ganga Project Directorate (GPD) in the Ministry of Environment and Forests for regeneration and development work on this major river basin of country. The Authority, under the Chairmanship of the Prime Minister reviews the overall progress of Ganga Action Plan at regular intervals and provides guidance. The Monitoring Committee and the Steering Committee of the Action Plan also review the progress from time to time. The three State Government of U. P., Bihar and W. Bengal are involved in the plan. A number of schemes (262) necessary for plan have already been sanctioned to these three states costing ₹ 258.44 crores. The major objective of the Action Plan is the immediate reduction of pollution load on the river and the establishment of self sustaining treatment plant system. Keeping in these views, following components have been identified as the components of the action plan:

1. Sewer system renovation.
2. Diversion of sewer from/ through interception
3. Renovation and construction of new and old sewage treatment plant, pumping stations etc.
4. Appropriate arrangement for sewer and waste utilization
5. Diversion of sewer lines to nearest treatment plants for treatment.
6. Prevention of pollution through river bank communities and cattle population.
7. Launching of low cost sanitation to reduce pollution load.
8. Application of various biological methods for stream purification.
9. Establishment of pilot projects for resource energy recovery.
10. Control dead body disposal in the river.

The action plan also laid down in detail the names of cities and towns situated on the Ganga, their population, quantity of sewage generated, cost of installation, maintenance and operation of sewage pumping and treatment plants and revenues to be generated from resources recycling units.

Evaluation of ganga action plan: A comprehensive evaluation of GaNGA Action Plan by independent agencies was undertaken in April 1995. A cost benefit analysis was under taken with the assistance of the Department of for International Development of UK. The conclusions and recommendations of these studies are being used to bring about improvement in the subsquent Ganga Action Plan Phase- II and the National River Conservation Plan.

(*h*) **National river conservation plan including ganga action plan phase-II** : Through a Government Resolution dated 5-12-1996 GAP Phase-II has been merged with the National River Conservation Plan (NCP). Thus the expended NRCP covers 141 towns located along 22 inter- State rivers in 14 States. The total cost of the scheme is ₹ 2013.40 crores. During the 8th Five year Plan, NRCD and GAP Phase-II were Centrally Sponsored Schemes with 50:50 cost sharing between the central government and the concerned State governments. The NRCA at its IX meeting held on 12-07-97 decided to convert NRCP in to 100% funded Centrally Sponsered Scheme. The Cabinet Committee on Economic affairs approved the 1000% funding pattern in 1998.

(*i*) Yamuna action plan: Under this action plan , pollution abatement works ar being taken up in 21 towns. Of these 12 are in Haryna, 8 in Uttar Pradesh desides Delhi. So far 31 schemes have been completed. Out of the project outlay of 496.45 crore the expenditure incurred totals to 390.84 crore . External assistance of Yen 17.77 billion is being provided by OECF, Japan to partly finance the Yamuna Action Plan.

(*j*) Gomti action plan: Under this component, pollution abatement works ae being taken up in Lucknow, Sultanpur and Jaunpur in Uttar Pradesh. About 269 mld of sewage is targeted to be intercepted, diverted and treated under this action plan.

(*k*) Damodar action plan: Under this Action Plan, pollution abatement works are being taken up in 12 towns. Of these 8 are in Bihar and 4 are in West Bengal. However, due to low priority given by West Bengal government and problems of operation and maintenance of GAP I assets in Bihar, work on DHP has not progressed.

(*l*) Control of Hazardous substances: The Environmental (Protection) Act, 1986 specifically defines the hazardous substances in Indian legal environment. These substances may have the property to produce toxic effects, are highly reactive and flammable in nature. Due to such nature, special attention is required during handling, transportation etc. The common methods include:

1. Substitution *i.e.* changing to less hazardous substances from chronic substance.
2. Change/ modification in production/ manufacturing/ handling processes.
3. Isolation of the system.
4. Use of wet processes instead dry processes.
5. Efficient use of ventilation/ exhaust etc. with proper arrangement of exhaust vents etc.
6. Dilution method.
7. Application of personal safety devices/ equipments.
8. Personal hygienic habits.
9. Good house keeping and their maintenance.
10. Disposal of waste material.
11. Special control procedures.

12. Periodic medical examinations and control
13. Training and education of the personnel concerned to the process/ handling etc.

(*m*) Notification on Uniform Consent Procedure to be followed by State Pollution Control Boards and Pollution Control Committees: A notification containing Draft Rules to provide for uniform consent procedure to be followed by the SPCBs and PCCs has been prepared and notified on 20th December 1999 under the Environment (Protection) Act, 1986. The objective of the notification is to bring out rationalization in the consent management practices being followed by the Pollution Control Boards in the State and Union Territories for the purpose of Water (Prevention and Control) Act, 1974, the Air (Prevention and Control of Pollution) Act, 1981 and Authorization Under the Hazardous Waste (Management and Handling) Rules 1989.

CHAPTER 2

Air Pollution

Our atmosphere has received smoke and other pollutants for millions of years from many natural processes such as volcanoes, natural fires and dust storms. But the biosphere does have means for removing, assimilating and recycling these natural pollutants. They disperse and dilute in the atmosphere. Then they settle or come down to earth via precipitation, and poisonous gases are converted to harmless products by soil micro-organisms. Thus keeping the natural inputs of pollution well below threshold level. However, the human inputs of pollutants are overloading the natural system and it has been calculated that human sources exceed those from natural sources by 10 to 1000 fold, depending on the particular pollutant.

Earth's atmosphere: The earth is the only planet in the solar system and in the universe, known so far, to have the unique atmosphere where life can sustain. Its distance from the sun and its atmosphere makes its environment warm and inhabitable. It is the only planet where water exists in all three forms. The composition of the earth' s atmosphere is determined by biological processes together with physical and chemical changes. Temperature profile is used to distinguish different regions of the atmosphere in terms of layer characteristics and by specific vertical temperature gradients. It has thus been divided into the following layers:

Troposphere: From the earths surface, the mean temperature decreases with height up to about 10 to 16 km. This region is known as the troposphere. The boundary where temperature decrease ceases is known as the tropopause. The general distribution of temperature at the tropopause is such that a minimum value is found at the equator and a maximum value in the polar regions. The troposphere acts as a source of heat resulting from the absorption of the visible sun light. It is characterized by its negative temperature gradient of about 6 degree K/km.

Stratosphere: The tropopause marks the end of troposphere and the beginning of the stratosphere, the atmosphere's second layer, which extends from about 17 to 48 km above the earth's surface. Although the stratosphere contains less matter than the troposphere, its composition is similar with two notable exceptions: its volume of water vapor is about 1000 times less and its volume of ozone is about 1000 times greater. The presence of ozone in the stratosphere keeps about 99% of the suns harmful UV radiations from reaching the earth's surface.

Mesosphere: The region between stratopause and about 85-90 km is known as the mesosphere and its upper boundary is called mesopause. The mesopause is the coolest region of the atmosphere.

Thermosphere: Above the mesopause, the region is known as the thermosphere. In this region, the temperature increases to reach highest value. The high temperature in the thermosphere are caused by very thin atmosphere.

There is no boundary between the atmosphere and void of outer space. About 75% of the earth's atmosphere lies within 16 km. of the surface and 99% of the atmosphere lies below an altitude of 30 Km. The atmosphere is an insulating blanket around the earth. It is sources of essential gases, maintains narrow difference of day and night temperatures and provides a medium for long-distance radio communication. It also acts as shield around the earth against lethal UV radiation's and meteors. Without atmosphere, there will be no lightening, no wind, no clouds, nor rains, no snow and no fire. Normal composition of clean air at or near sea is as given below :

Gases	Percent (by volume)
Nitrogen	78.084
Oxygen	20.9476
Argon	0.934
Carbon dioxide	0.0314
Methane	0.0002
Hydrogen	0.00005
Other gases	minute

However, it is very difficult to find such clean air and the concentration levels of different gases show variations due to addition of different types of pollutants in to the air which may be in the form of gases or particulates making it unfit for living organisms. In the western and more developed countries, it was realized in the 1950s and 60s that unrestricted discharge of pollutants in to the atmosphere could no longer be tolerated. In Indian cities, particularly in the metros, the layers of dust on almost all fixtures of the house, cleaned only a few hours back; the pungent, turbid air in areas of dense vehicular traffic; burning, watering eyes after a scooter ride are some examples which remind us that we may no longer be silent spectators to a situation where the environment is actually hitting back at us. Residents of our metropolitan cities are today exposed to some of the highest levels of air pollution in the country and perhaps in the world.

AIR POLLUTION IN INDIA

Industrialization and urbanization have resulted in a profound deterioration of India's air quality. Of the 3 million premature deaths in the world that occur each year due to outdoor and indoor air pollution, the highest number are assessed to occur in India. According to the World Health Organization, the capital city of New Delhi is one of the top ten most polluted cities in the world. Surveys indicate that in New Delhi the incidence of respiratory diseases due to air pollution is about 12 times the national average.

According to another study, while India's gross domestic product has increased 2.5 times over the past two decades; vehicular pollution has increased eight times, while pollution from industries has quadrupled. Sources of air pollution, India's most severe environmental problem, come in several forms, including vehicular emissions and untreated industrial smoke. Apart from rapid industrialization, urbanization has resulted in the emergence of industrial centers without a corresponding growth in civic amenities and pollution control mechanism.

Regulatory reforms aimed at improving the air pollution problem in cities such as New Delhi have been quite difficult to implement, however. For example, India's Supreme Court recently lifted a ruling that it imposed two years ago which required all public transport vehicles in New Delhi to switch to compressed natural gas (CNG) engines by April 1, 2001. This ruling, however, led to the disappearance of some 15,000 taxis and 10,000 buses from the city, creating public protests, riots, and widespread "commuter chaos." The court was similarly unsuccessful in 2000, when it attempted to ban all public vehicles that were more than 15 years old and ordered the introduction of unleaded gasoline and CNG. India's high concentration of pollution is not due to a lack of effort in building a sound environmental legal regime, but rather to a lack of enforcement at the local level. Efforts are currently underway to change this as new specifications are being adopted for auto emissions, which currently account for approximately 70% of air pollution. In the absence of coordinated government efforts, including stricter enforcement, this figure is likely to rise in the coming years due to the sheer increase in vehicle ownership. (Source: US Energy Information Agency).

Sources of Air Pollutants and their Effects

In large measures, air pollutants are direct and indirect byproducts of burning coal, gasoline and other liquid fuels and refuse. These fuels and wastes are organic and with complete combustion, the byproducts are CO_2, and water vapor. However, their complete oxidation rarely takes place. In addition, fuels or refuse contain impurities or additives and these are also emitted into the air when burned. Some of these direct products further react within the atmosphere and produce additional indirect products. We may say that air pollution results from gaseous emissions from mainly industry, thermal power stations, automobiles, domestic combustion etc.

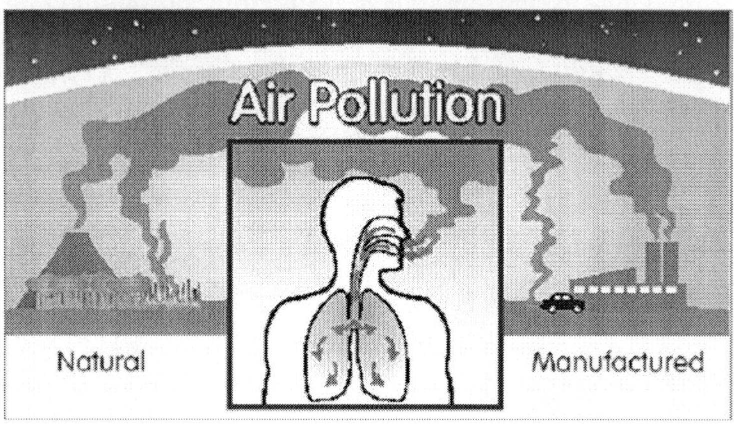

A. Industrial emissions : There are a number of industries which are source of air pollution. Petroleum refineries are the major source of gaseous pollutants. The chief gases are SO_2 and NO_x. The fertilizer industries release oxides of nitrogen and dust particles of microscopic size. Dust particles may be evolved from the process of drying, burning, calcining, grinding, screening, mixing and packaging. Cement dust is a common air pollutant around cement factories and construction sites which emit plenty of dust, which is potential heath hazard. Chemically it is a mixture of oxides of aluminum, silica, potassium, calcium and sodium. Stone crushers and hot mix plants also create a menace. The SPM levels in such areas of stone crushing are more than five times the industrial safety limits. There are many food and fertilizers industries, which emit gaseous pollutants. Many chemical manufacturing industries emit acid vapors in air. In the past two decades we have seen a tremendous increase in the number of industries in our metros particularly in Delhi, Bombay , Calcutta and Bangalore. Many of these are located in predominantly residential areas. The industries producing sulphuric acid, discharge large quantities of sulfur dioxide in the air and pollute the atmosphere up to a few kilometers. The industries producing fluoride compounds have serious effects on plants, animals and human beings. The toxic effects of fluorides on live stocks arise from ingesting contaminated forage on which fluoride dust has settled. The various iron and steel industries emit sulfur dioxide, oxides of metals and carbon mono and di oxides. Acids are also produced in the atmosphere due to reaction of sulpher di-oxide and nitrogen oxides with water vapors.

B. Automobiles : Starting from the 1950s, when there was a mushrooming use of cars for commuting, the metropolitan areas increasingly became enshrouded in a brownish haze on a daily basis. This haze is called smog, or more properly, photochemical smog, because sun light is involved in its formation. Certain weather conditions intensify the smog levels, the most important being temperature inversions. Normally, air temperature decreases with increasing elevation. In this situation, the warm air near the ground rises, carrying pollutants upward and dispersing them at higher altitudes. In a temperature inversion, a layer of cold air near the ground is covered by warm air. This situation develops as an influx of cooler air moves in under the warm air. With a temperature inversion, the upward movement of air carrying pollutants is blocked and pollutants accumulate in the cool air near the ground. The toxic vehicular exhausts are sources of considerable air pollution, next only to thermal power plants. The ever-increasing vehicular traffic density has posed a continues threat to the ambient air quality. There are over 300 million cars, trucks and buses in the world over and their number increasing rapidly. India, which is likely to have over 30 million vehicles, have more than 65% two-wheelers operating on petrol. In all the major cities of the country about 800 to 1000 tones of pollutants are being emitted into the air daily, of which 50% come from automobiles exhausts. In the major metropolitan cities, vehicular exhaust accounts for 70% of all CO, 50% of all hydrocarbons, 30-40%of all oxides and 30% of all SPM. In cities like Delhi, during peak traffic hours, automobiles of all classes emit as much as 700 Kg of CO, 250 Kg of hydrocarbons and 60 Kg of nitrogen oxides. The two –wheelers and three-wheelers contribute 60% of the total CO, and 83% of total hydrocarbons, whereas heavy traffic vehicles

55 to 80% of the oxides of nitrogen. It is estimated that a car (without cleaning device) on burning 1000 liter of fuel emits 350 Kg CO, 0.6 Kg SO_2, 0.1 Kg lead and 1.5 Kg SPM.

The sources of emission in automobiles are: (*i*) exhaust system, (*ii*) fuel tank and carburetor and (*iii*) crankcase. The exhaust produces many air pollutants including unburnt hyrocarbons, CO, NO_x lead oxides. There are also traces of aldehydes, esters, ethers, peroxides and ketones which are chemically active and combine to form smog in presence of light. Evaporation from fuel tank goes on constantly due to volatile nature of petrol, causing emission of hydrocarbons. The evaporation through carburetor occurs when engine is stopped and heat builds up, and as much as 12 to 40 ml of fuel is lost during each long stop causing emission of hydrocarbons. Some gas vapor escapes between walls and the piston, which enters the crankcase and then discharges into the atmosphere. This accounts for 25% of the total hydrocarbon emission of an engine.

C. Nuclear and thermal power stations: The catastrophic accident at Chernobyl in USSR in April 1986 is the biggest example of pollution that is being caused by thermal power stations. Radioactive emissions may penetrate through biological tissues in a manner analogous to tiny bullets. They leave no visible marks nor are they felt, but they are capable of breaking molecules within cells. In high doses, radiation may cause enough damage to prevent cell division. In low doses it may damage to DNA molecules. There are a number of thermal power stations and super thermal stations in the country. The National Thermal Power Corporation (NTPC) has setup mammoth coal-powered power stations to augment the energy generation of our country. The coal consumption of thermal plants is several million tones. The chief pollutants are fly ash, SO_2 and other gases hydrocarbons.

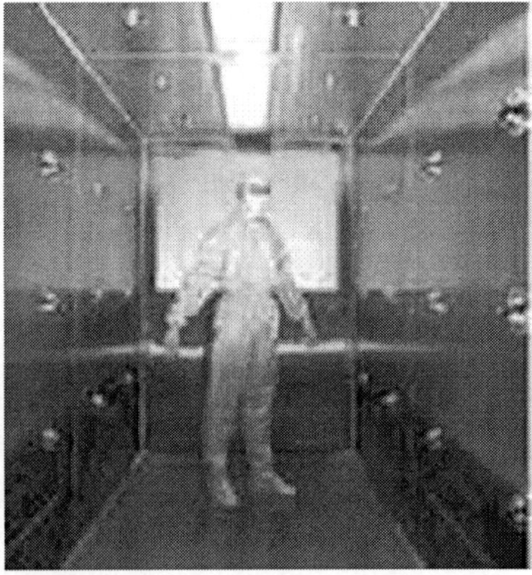

Fig. 2.1 : Working safely in nuclear power stations

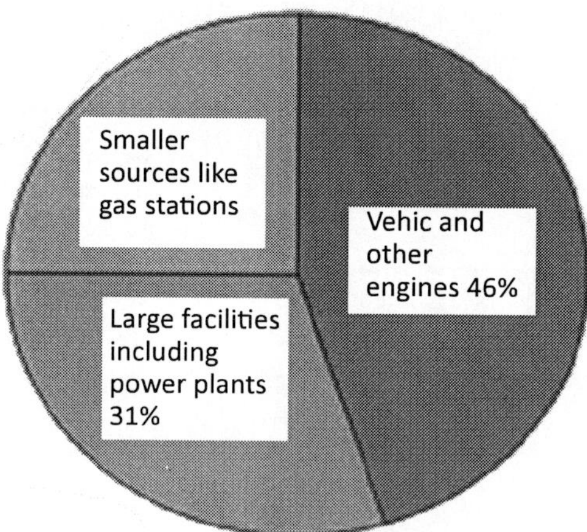

Fig. 2.2 : Percent contribution of different air pollutants

What happens to pollutants in the atmosphere: Once the pollutants enter the troposphere, they are affected by four processes:

1. Transport, in which pollutants are carried down wind. The extent of this process depends on wind speed and direction, topographical features and how high in the atmosphere pollutants are injected.
2. Dilution, in which the turbulent motion of air mix and reduce the concentration of pollutants.
3. Transformation, in which pollutants undergo physical changes, including agglomeration, chemical changes and photochemical changes triggered by reaction with incoming solar energy.
4. Removal from the atmosphere by rainout, washout and fallout.

Choking of south asia by a blanket of smog: According to a report published in Times of India dated 13th August, 2004, India and rest of the South Asia is covered by a deadly, three km. deep blanket of pollution, which is readily changing monsoon patterns, causing draughts and literally killing hundreds of thousands of people by respiratory diseases. A group of scientists working with the United Nations Environment Program on the Asian Brown Haze has released this study. It said that the vast pollution parcel could endanger the economic success of South Asian Countries, particularly India. The haze has been identified as a deadly cocktail of ash, acids, aerosols and other particles. The study also says that the worst may be yet to come and the effects of the haze would intensify over the next 30 years. Already, higher levels of respiratory diseases are leading to several hundreds of thousands of premature deaths, as revealed by data from seven Indian cities, including Ahmedabad, Kolkatta, Delhi and Mumbai. The Indian data suggested some kind of air pollution was responsible for 24,000 annual premature deaths in the early 1990s, said the report. A few year later, the number rose to an estimated 37,000 per year.

UNEP executive director, Klaus Toepfer says, "In India they are expecting more than two million people to die because of the incomplete burning of biomass." The open fires used by the majority of ordinary Indians to cook their food also contribute to this. This pollution parcel, which stretches three kilometers high, can travel half way round the globe in a week, thus having global implications.

AIR POLLUTANTS AND THEIR EFFECTS

Some of the air pollutants and their effects have been discussed here below:

(*a*) Compounds of Carbon: The two important pollutants are carbon dioxide and carbon monoxide.

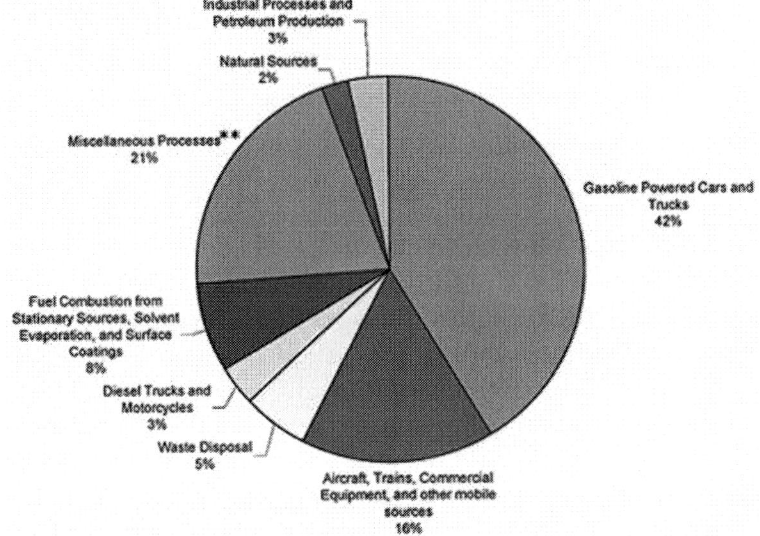

1. Carbon dioxide : Human activities are responsible for the increasing amount of this gas in our atmosphere The concentration of CO_2 in the earth's atmosphere currently is 360 μl/L, which is approximately 30% higher than the pre industrialization level and is conservatively projected to double by the end of first decade of 21 centuary. Around 1900, the concentration of CO_2 in air was 290 ppm, in 1960 it was about 315 ppm and it is increasing at a rate of approximately 1.0 mg/kg/year. The reasons given for this situation are: (*a*) depletion of forest cover, (*b*) burning of fossil fuel, (*c*) increased number of living organisms and, (*d*) upset of balance in exchange of CO_2 between oceans and atmosphere.

Major amount of carbon dioxide is released in the atmosphere from burning of fossil fuel (coal, oil etc) for domestic cooking, heating etc. and the fuels consumed in furnaces of power plants, industries, hot-mix plants etc. From fossil fuel alone more than 18×10^{12} tones of CO_2 is being released into atmosphere each year. In our country, on an average, thermal power plants are likely to release around 50 million tones of CO_2 each year in the atmosphere.

To some extent an increase in CO_2 level in atmosphere increases the photosynthesis rate and consequently plant growth, acting as fertilizer especially in hot tropical climates. This potential of fertilizer effect may be exploited by using modified crop varieties and agricultural practices. However, an increase in CO_2 concentration in atmosphere may result into disastrous effects also.

Greenhouse effect: The term green house effect was coined by Jean Fourier more than 100 years ago. Since CO_2 is confined exclusively to the troposphere, its higher concentration may act a serious pollutant. Under normal conditions (with normal CO_2 concentration) the temperature at the surface of the earth is maintained by the energy balance of the sun rays that strike the planet and heat that is radiated back into space. The earth's surface and atmosphere receive energy from the sun. This energy comes in the form of radiation mainly in the visible region, 1 = 400 – 700 nm. Much of UV light (1 < 400 nm) from the sun is filtered out in the stratosphere and warms the air there rather than at surface of earth. Besides visible light, we receive some infra red (IR) light in the region 800 – 4000 nm. Of the total incoming light that impinges upon the earth, about 50% reaches the surfaces and is absorbed by it. About 20% of the incoming light is absorbed by gases in the atmosphere, such as, UV light by ozone in stratosphere and IR by CO_2 and water vapor in air. The remaining 30% is reflected back in to space by clouds, ice, snow and other reflecting bodies without being absorbed.

Like any other warm body, the earth emits energy, the amount of the energy that the planet absorbs and the amount that it releases must be equal if the temperature is to remain constant at earth. The emitted energy is neither visible nor UV, it is IR light having wavelength in the range of 4000-5000nm which is called thermal infra red region.

Some gases in the air such as CO_2, temporarily absorb thermal infra red light and, therefore, not all IR emitted from the earth's surface and atmosphere escape directly in to space. Shortly after its absorption by CO_2 molecules, this thermal IR light is remitted in all directions. Thus some thermal IR is redirected back towards the earth's surface and is reabsorbed, and consequently further heats to both, the surface and the air. This phenomenon of redirection of thermal IR towards the earth is called the green house effect. Thus, the green house effect is trapping of heat caused by gases such as carbon di-oxide and water vapor which are transparent to incoming solar radiations but re-emit the infra red radiations from earth's surface. This is responsible for keeping the temperature of earth's surface at +15° C rather than –15° C (if there would had been no atmosphere).

The phenomenon that worries us is that, increasing the concentration of trace gases in atmosphere, that absorb thermal IR, would result in the redirection of even more of the out going thermal IR energy and would increase the average temperature of earth surface beyond+15° C. Most of the heat is absorbed by CO_2 layer and waters vapors in the atmosphere, which adds to the heat that is already present. CO_2 increases the earth temperature by 50% while CFCs are responsible for another 20% increase and there are enough CFCs up there to last in 120 years.

The heat trap provided by atmospheric CO_2 probably helped to create the conditions necessary for the evolution of life and the greening of earth. Compared to moderately warm planet, Mars, with too little CO_2 in its atmosphere, is frozen cold and Venus with too much is a dry furnace. The excess CO_2 to some extent is absorbed by the oceans. But with the industrialization of West and increased consumption of energy, CO_2 was released into atmosphere at a faster rate than the capacity of oceans to absorb it. Thus its concentration increased. According to some estimates CO_2 in air may have risen by 30% since the middle of 19th century. It may even be doubled by 2030 A. D.

There are some difference of opinions, however, about the extent of rise in earth's temperature due to increasing CO_2 levels. According to some computerized models, doubling the CO_2 level will increase the global mean temperature (15°C) by 2 °C. But, some others say that this will be less than one quarter of a degree. However, even a change of two degrees may disrupt the earth's heat budget, causing catastrophic consequences.

The green house effect is being greatly intensified by certain other gases that human beings are introducing into the atmosphere. In particular, methane, chloroflurocarbons (CFCs), and nitrous oxides absorb infra-red 50 to 100 fold more effectively than carbon di oxide. Consequently, while the amounts of these gases being emitted into the atmosphere are far less than the amounts of carbon di oxide, they have an effect on climatic warming which nearly equals that caused by the CO_2 additions. The term green house gases is used to denote CO_2 and these other gases that absorb infra red radiations and lead to climatic warming.

How to cope with the greenhouse effect: Following are the ways by which we can lessen the addition of green house gases in the atmosphere and bring a sustainable balance :

(*i*) By increasing the fuel efficiency of vehicles and all other forms of energy conservation since energy production is nearly all based on the burning of fossil fuels.

(*ii*) Use of non-conventional sources of energy and develop and implement no-fossil fuel energy alternatives.

(*iii*) To put a check on clearing of forests, particularly those of tropical regions.

(*iv*) By carrying out maximum plantation of trees.

Impacts of enhanced green house effect: The enhanced green house effect will not only cause global warming but will affect various other climatic and natural processes.

1. **Increase in global temperature**: The global temperature has been observed to increase ever since systematic recording of this temperature

has been initiated. The world temperature has increased by 0.7°C from 1860 to 1990AD. Where as temperature of tropical ocean has risen by 0.5°C for the last 30 to 50 years. It is estimated that the earth's mean temperature will rise between 1.5 t0 5.5 °C by 2050 if input of green house gases continue to rise at present rate.

2. **Climate change:** The global change in temperature will not be uniform everywhere and will fluctuate in different regions. The places in higher latitudes will be warmed up more during late autumn and winter than the places in tropics. Poles may experience 2 to 3 times more warming than the global average while warming in the tropics may be only 50 to 100% of the average. The increased warming at poles will reduce the thermal gradient between the equator and high latitude regions decreasing the energy available to the heat engine that drives the global weather machine. This will disturb the global pattern of winds and ocean currents as well as timings and distribution of rainfall. Disturbed rainfall will result in some areas becoming wetter and the others drier.
3. **Rise in sea level:** With the increase in global temperature sea water will expand. Heating will melt the polar ice sheet and glaciers resulting in further rise in sea level. Current modes indicate that an increase in the average atmospheric temperature of 3°C would raise the average global sea level by 0.2 to 1.5 m over the next 50-100 years. One meter rise in sea level will inundate low lying areas of cities like Sanghai, Cairo, Bangkok, Sydney etc. Life of millions of people will be affected by the sea level rise.
4. **Effects on human health:** The global warming will affect the distribution of vector of disease like mosquitoes. The diseases like malaria, filariasis and elephantiasis will increase. More temperature and humidity will increase respiratory and skin diseases.
5. **Effects on agriculture:** Tropical and subtropical regions will be affected more since the average temperature in these region is already on higher side. Even a rise of 2°C may be quite harmful to crops. Soil moisture will decrease, evapo-transpiration will increase which may drastically affect many crops.

Kyto protocol: Over a decade ago, most countries joined an international treaty- the United Nations Framework Convention on Climate Change- to consider what could be done to reduce global warming and cope with whatever temperature increase are inevitable. In 1997, the governments agreed to an addition to the treaty, called the Kyoto Protocol, which legally bound industrialized nations to reduce world wide emmission of green house gases by an average of 5.2% below their 1990 level by 2012. The cut aren't uniform-while most of the Western European Countries committed to cuts in the vicinity of

85, the rates vary across countries. Some countries such as Iceland, can actually increase emmission upto 10% under the treaty, which came in to effect on February 16, 2005.

The treaty was signed by 141 nations, which together account for 55% of green house gas emmissions globally. Significantly, the US and the Australia haven't signed the treaty. In 2001, one of the first act of then newly elected President of USA was to formally withdraw the US from Kyoto. Bush said that the US would not ratify the treaty because it would damage the American economy and major developing nations like China and India were not covered by its provisions. Althgough India and China have signed the treaty, they aren't required to make any cuts in emmissions till 2012, the logic being that the developing nations shouldn't pay a price for late industrialization.

The Kyoto protocol contains measures to assess performance and progress. It also contains some penalties. Countries that fail to meet their emmission target by the end of first commitment period (2012) must makeup the difference plus a penalty of 30% in the second commitment period. Their ability to sell credits under emmissions trading will also be suspended.

Is the world's climate really changing: Many Scientists believe so. The UN's Intergovernmental Pennal on Climate Change (IPCC) summarises the work of 2,000 of the world's top most climate experts. Its reports say that the world is definitely getting warmer. The IPCC says the average global surface temperature has risen by about 0.6 degrees Celsius since 1900, with much of that rise coming in the 1990s- probably the warmest decade in 1,000 years. The IPCC also found that the snow cover since the late 1960s has decreased by 10% and lakes and rivers in the Northern Hemisphere are frozen over about two weeks less each year than they were then. Mountain glaciers in non-polar regions have also been in noticeble retreat in the 20th centuary, and the average global sea level has risen between 0.1 and 0.2 meters since 1900.

Albido: Albido is a Latin word meaning-whiteness. It is defined as" Albido is the ratio of the total radient energy flux reflected or scattered in all directions by an object to the total radiant flux intercepted by the object. This formal definition is strictly appropriate only in the context of certain special cases. These concern reflection of sunlight from entire and isolated bodies such as planets, or the reflection of sunlight from individual particles or molecules.

Ground level measurements of albedo are not made in a way that is consistant with the above definition. Using a horizontal surface as an example, it's measured albedo is the ratio of the upward component of the flux density (flux per unit area) of radiation scattered or reflected from the surface to the downward component of the incoming flux density. The measurement is normally taken in the context of solar radiation reflected by the ground or by some other horozontal surface such as a deck of cloud. It is obtained by an albedometer, which measures separately and simultaneously both the up-welling and down-welling flux densities at some small distances above the surface in question. These sensors are flat-plate absorbers. Thus, the typical observation of albedo does not involve measurement of radiation scattered in all directions by the particular small area of surface. Rather it involves measurement of the radition scattered into a horizontal unit area above the surface from all parts of the surface within view of the downward looking flat plate sensor.

Albedo is not an intrinsic single-valued property of a surface, but is normally a function of the manner in which the surface is illuminated. In cloud free skies where the direct beam of Sun dominates the down coming solar radiation, the albedo of virtually all ground

2. Carbon monoxide : It is produced by the incomplete combustion of coal, charcoal, petrol and its products. The largest source of combustion producing CO are internal combustion engines. From combustion sources, the CO emission per year in the world is about 2.6×10^8 tones. Furnaces, stoves, open fires, factories, power plants give off CO. The source of atmospheric CO is the oxidation of methane. Anaerobic decomposition of organic matter produces methane, which is oxidised to produce CO. Photochemical decomposition of ozone and the reaction of atomic oxygen with water vapor produces hydroxyl radicals, which initiate oxidation of methane. Thus CO production in atmosphere is associated with photochemical reactions.

Carbon mono oxide combines with hemoglobin of blood, reducing its O_2 carrying capacity. The gas is fatal over 1000 ppm, causing unconsciousness in an hour and death in four hours. If this gas is inhaled for few hours at even a low concentration of 200 ppm, it causes symptoms of poisoning. Inhaled CO combines with blood hemoglobin to form **carboxyhemoglobin** about 210 times faster than O_2 does. Formation of carboxyhemoglobin decreases the overall O_2 carrying capacity of blood to cells resulting into oxygen deficiency-**hypoxia**. At about 200 ppm concentration for 6-8 hours, headache begins, and mental activity gets reduced; above 300 ppm, throbbing headache starts followed by vomiting and collapse; at 500 ppm, man reaches into coma and at 1000 ppm, is death occurs. The accepted maximum allowable concentration (MAC) for occupational exposure is 50 ppm for 8 hours. The increase in carboxyhemoglobin level form 1-2% to 3-4% may cause cerebral anoxia resulting into impairing of vision and psychomotor activity. Sub-lethal concentrations of this gas may be injurious due to prolonged exposure. In smokers, prolonged exposures may cause an adaptive response, even producing more hemoglobin, as high as 8%. At 10% carboxyhemoglobin in blood due to smoking there may be lowered tolerance to CO. Cigarette smokers have increased hematocrit (per cent volume of red blood cells), within minutes of smoking. In developed countries cigarettes are linked to at least 80% of all deaths from lung cancer. Most plants are not affected by CO levels known to affect man. At higher levels (100 to 10,000 ppm), the gas affects leaf drop, leaf curling, reduction in leaf size, premature aging etc. It inhibits cellular respiration in plants.

3. Compounds of sulphur: Sulpher remains in the atmosphere in three forms namely, sulphur dioxide, hydrogen sulphide and sulphate particles.. Sulphur dioxide is produced mainly by the combustion of the fossil fuels like coal, mobile oil and petrol. The other S-compounds are carbonyl sulphide (COS), carbon disulphide (CS_2), dimethyl sulphide [$(CH_2)_2$ S] and sulphates.

(*a*) Sulphur dioxide: It is a colorless gas with a suffocating and pungent odor. The major source of SO_2 emission are industrial processes and burning of sulpher containing coals in thermal power plants, smelting industries (smelting sulphur containing metal ores) and other processes as manufacturing of sulphuric acid and fertilizers. These account for about 75% of the total SO_2 emission. Most of the rest 25% emission is from petroleum refineries and automobiles. In U. S. A., in 1970 there was emitted 37 million tons of $SO_{2.}$ It is believed that about 109 million tons of SO_2 are added each year into the global environment. In polluted areas its concentration may reach 3000 mg/m^3, but in non polluted areas the concentration usually remains 0.3 to 1.0 mg/m$^{3.}$ In our country, SO_2 emission is on the increase over the years and it has almost doubled between 1979 and 2000 A.D. The life time of SO_2 in the atmosphere depends upon the rate at which it is oxidized On global scale, about half of the ambient SO_2 originates from the oxidation of hydrogen sulphide given off by decaying organic matter.

The chronic exposures to SO_2 causes intense irritation to eyes and respiratory tract. It is absorbed in the moist passage of upper respiratory tract, leading to swelling and stimulated mucus secretion. Exposure to 1 ppm level of SO_2 causes a constriction of the air passage, and causes significant bronocho-constriction in asthmatics even at a lower (0.05-0.50 ppm) concentrations. Moist air and fogs increase the SO_2 dangers due to formation of H_2SO_4 and sulphate ions; H_2SO_4 is a strong irritant (4-20 times) than SO_2.

Plants are affected at lower concentrations of SO_2 than human beings. Leaf bleaching, necroses and loss of yield are the main damages caused by $SO_{2.}$ Growth of the ornamental plants *Calendula officinalis* and *Dahlia rosea* was adversely affected when exposed to 1to 2 ppm of SO_2 for two hours.

SO_2 is also involved in the erosion of building materials as limestone marble, the slate used in roofing, mortar and deterioration of statues. Petroleum refineries, smellers, Kraft mills deteriorate the adjoining historic monuments.

(*b*) Hydrogen sulphide: The chief sources of H_2S are decaying vegetation and animal matter, especially in aquatic habitats. Sulphur springs volcanic eruptions, coal pits and sewers also give off this gas. About 30 million tons of H_2S every year is released by oceans and 60 to 80 million tons per year by land. Industries emit about 3 million tons every year. The chief industrial sources of H_2S are, users of sulphur containing fuel. In polluted areas its concentration reaches 100mg/m^3 , while in unpolluted areas its concentration remains around

0.3 mg/m^3. It is generally believed that hydrogen sulphide is oxidised in to sulpher dioxide but there is not much evidence for this.

At a low concentration, H_2S causes headache, nausea, collapse, coma and finally death. Unpleasant odor may destroy the appetite at 5 ppm level in some people. A concentration of 250 ppm may cause conjunctivitis and irritation of mucus membranes. Exposure at 500 ppm for 15-30 minutes may cause colic diarrhea and bronchial pneumonia. This gas readily passes through alveolar membrane of the lung and penetrates the blood stream. Death occurs due to respiratory failures.

4. Nitrogen Oxides (NO_x): Nitrogen oxides are major primary pollutants. There are measurable amounts of Nitric oxide, nitrous oxide and nitrogen dioxide present in the unpolluted atmosphere. Of these nitric oxide (NO) is the pivot compound. It is produced by combustion of O_2 and N_2 during lighting discharges and by bacterial oxidation of NH_3 in soil. Nitric oxide contacts with air and combines with O_2 or even more readily with O_3 to form the more poisonous nitrogen dioxide (NO_2). NO_2 may react with water vapor in air to form HNO_3. This acid combines with NH_3 to form ammonium nitrate. Fossil fuel combustion also contributes to oxides of nitrogen. About 95% of the nitrogen oxide is emitted as NO and remaining 5% as NO_2. In urban areas about 46% of oxides of nitrogen in air comes from vehicles and 35% from electric generation and the rest from other sources.

(*i*) **Nitrous oxide (N_2O):** This gas has so far not been implicated in air pollution problems. The concentration of N_2O remains more or less constant up to the troposphere and then decreases with increase in altitude.

(*ii*) **Nitric oxide (NO):** The chief source of this gas are the industries manufacturing HNO_3 and other chemicals, and the automobile exhausts. At high temperature, combustion of gasoline produces this gas. A large amount of this is readily converted in to more toxic NO_2 in the atmosphere by a series of chemical reactions. NO is responsible for several photochemical reactions in the atmosphere, particularly in the formation of several secondary pollutants like PAN, O_3, carbonyl compounds etc. in the presence of other organic substances.

(*iii*) **Nitrogen dioxide (NO_2):** It is a deep reddish brown gas, which is the only widely prevalent colored pollutant gas. This gas is the chief constituent of **photochemical smog** in metropolitan areas.

At lower concentrations plant absorb NO and NO_2 from air without any damage. The uptake rate of NO_2 is much greater than that of NO. Concentrations of NO_2 in access of 2 mg/kg of air causes leaf damage to sensitive plants. In some plants, photosynthetic activity is impaired at a concentration of 0.6 mg/kg. In high concentrations, the oxides of nitrogen cause irritation to the mucus membrane and damage to the respiratory system.

5. Photochemical products: The important photo chemical products which are produced in the atmosphere as a result of interlinking of oxides of nitrogen, hydrocarbons and ozone in the presence of light include Olefins, aldehydes, ozone, PAN and photochemical smog. Olefins are produced directly form the exhaust and in the atmosphere from ethylene. At very low concentrations of few ppb, they affect plants seriously. They wither the sepals of orchid flowers, retard the opening of carnation flowers and may cause dropping of their petals. At high levels they retard the growth of tomatoes. Aldehydes as HCHO and olefin, acroleins irritate the skin, eyes and upper respiratory tract.

Photochemical smog: It was first discovered in Los Angeles, USA and is, therefore, also called Los Angeles Smog. When the air pollutants such as hydrocarbons and nitrogen oxides react with one another in the presence of sun light, NO_2, ozone and a compound called PAN (peroxyacetyl nitrate)- photochemical smog is formed which appears as a yellowish brown haze. The term smog is used for mixture of smoke and moisture in the air. The key reactant is NO_2. The important hydrocarbons for the production of this smog are olefins. The major changes that occur in atmosphere are that hydrocarbons and NO are discharged in to the atmosphere from automobile exhaust in large amounts in the peak activity hours of day. With increasing intensity of solar radiations, the concentration of NO goes on decreasing, the concentration of NO_2 increases and so does the amount of aldehydes. Then a decrease in NO_2 concentration later in the day accompanies the occurrence of significant ozone concentration, which from mid day to afternoon goes on increasing along with aldehydes and hydrocarbons. Similar photochemical reactions between oxides of nitrogen and hydrocarbons result in the formation of Peroxy Acetyl Nitrate (PAN).

Photochemical smog adversely affects plants, human health and materials. The oxidants enter as part of inhaled air, and alter, impair or interfere with respiratory process and other processes. Serious outbreaks of smog occurred in Tokyo, New York, Rome and Sydney in 1970, causing spread of disease as asthma and bronchitis in epidemic form. **Tokyo-Yokohama asthma** occurred in 1946 in some American soldiers and families living in smoggy atmosphere of Yokohama, Japan. Another serious disease caused by smog is **emphysema**, a disease due to structural breakdown of alveoli of lungs. The total surface area available for gaseous exchange is reduced and this cause severe breathlessness.

The smoke and the particulate matters (fog, mist, dust, soot etc) in the smog reduce the visibility, damage crops and livestock, and cause corrosion of metals, stones, building materials, painted surfaces, textile, paper, leather etc.

6. Particulate matter: This is a discrete mass of any material, except pure water, that exists as liquid or solid in the atmosphere and is of microscopic or submicroscopic dimensions. Air borne matter results not only from direct emission of particles but also from emissions of some gases that condense as particles directly or undergo transformation to form particle. Thus particulate matter may be primary or secondary. Primary particulate matter includes dust, as a result of wind, or smoke particles emitted from some factory, Atmospheric particulate matter range is size from 0.001 mm to several hundred mm. Particulate matter in atmosphere arises from natural as well as man-made sources. Natural sources are soil and rock debris (dust), volcanic emission, sea spray, forest fires and reactions between natural gas emission.

There are four types of sources of particulate matter: (*i*) fuel combustion and industrial operations (mining, smelting, polishing, furnaces and textiles, pesticides, fertilizers and chemical production), (*ii*) industrial fugitive processes (materials handling, loading and transfer operation), (*iii*) non-industrial fugitive processes (roadway dust, agricultural operations, construction, fire etc.) and (*iv*) transportation sources (vehicle exhaust and related particles from fire, clutch and break wear).

In our country, there is much of flyash introduced to atmosphere from fossil fuel based plants chiefly thermal power plants. They also emit coal dust. Besides them stone-crushers introduce smoke and dust in the atmosphere.

The particulate matter is injurious to health. Soot, lead particles form exhaust, assents, flyash, volcanic emission, pesticides, H_2SO_4, mist metallic dust, cotton and cement dust etc when inhaled by man cause respiratory diseases such as tuberculosis, and cancer.

In addition to the above there are also many kinds of biological particulate matter that remain suspended in atmosphere. These are bacterial cells, spores, fungal spores, and pollen grains. These cause bronchial, allergy and many other diseases in man, animals and plants.

7. Acid rains: The oxides of sulphur and nitrogen are important gaseous pollutants of air. These two react with water in the atmosphere producing sulphuric acid and nitric acid. These oxides are swept up into the atmosphere and can travel thousands of kilometers. The longer they stay in the atmosphere, the more likely they are to be oxidized into acids. These come down to the earth along with rain. The pH of acid rain may vary from 3 to 6. Natural rains are slightly acidic because rain water reacts with CO_2 in atmosphere and produces a weak acid, the carbonic acid. The acid rains contain sulphuric acid, nitric acid along with carbonic acid.

Effects of acid rains: (*i*) Acid rain makes the leaves of plants yellow and brown and accelerates senescence. Thus affecting the productivity of forests, grasslands and crops.

(*ii*) It affects the properties of soil, changes the distribution of organisms in it and causes damage to soil fertility.

(*iv*) It kills aquatic life forms and affects the productivity of aquatic ecosystems.

(*v*) It leads to respiratory and skin diseases.

(*vi*) It damages marble and lime stone monuments.

Ozone (O_3): In the atmosphere, ozone is largely found in the stratosphere. Ozone absorbs most of the UV radiation of the sun and thus acts as a shield, protecting the living organisms on earth from health hazards. The depletion of this O_3 layer by human activities may have serious implications and this has become a subject of much concern over the last few years. Ozone is also formed in the atmosphere through chemical reactions involving certain pollutants (SO_2, NO_2, aldehydes) on absorption of UV-radiations. The atmospheric ozone is now being regarded as potential danger to human health and crop growth.

The temperature decreases with increasing altitude in the troposphere (8 to16 km. from earth surface), while it increase with increasing altitude in the stratosphere (above 16 km. up to 50 km). This rise in temperature in stratosphere is caused by the ozone layer. The ozone layer has two important and interrelated effects. Firstly, it absorbs UV light and thus protects all life on earth from harmful effects of radiation. Second, by absorbing the UV radiation the ozone layer heats the stratosphere, causing temperature inversion which limits the vertical mixing of pollutants, thereby causing the dispersal of pollutants over the large areas near the earth surface. It is why a dense clouds usually hang over the atmosphere in highly industrialized areas causing several unpleasant effects. The wastes spread horizontally relatively fast (than slow mixing vertically), reaching all longitudes of the world in about a week and all latitudes within months. There is, therefore, very little that a country can do protect the ozone layer above it. The ozone problem is thus global in scope. In spite of slow vertical mixing, some of the pollutants (CFCs) enter the stratosphere and remain there for years until they are converted to other products or are transported back to the stratosphere. The stratosphere could be regarded as a sink, but unfortunately, these pollutants (CFEs) react with the ozone and deplete it.

The ozone near the earth's surface in the troposphere creates pollution problems. Ozone and other oxidants such as peroxy acetyl nitrate (PAN) and hydrogen peroxide are formed by light dependent reactions between NO_2 and hydrocarbons. Ozone may also be formed under UV-radiation effect. These pollutants cause photochemical smog. Increase in O_3 concentration near the earth's surface reduces crop yields significantly. It also has adverse effect on human health. Thus, while higher levels of O_3 in the atmosphere protects us, it is harmful when its comes in direct contact with us and plants at earth's surface.

In plants, O_3 enters through stomata. It produces visible damage to leaves, and thus a decrease in yield and quality of plant products. At 0.02 ppm it damages tobacco, tomato, bean, pine and other plants.

Ozone alone and in combination with other pollutants like SO_2 and NO_x is causing crop losses of over 50% in several European countries. In Denmark, O_3 affects potato, clover, spinach, alfalfa etc. In limited pockets O_3 concentration can be potentially harmful. For instance, in West Germany 100-250 mg/m^3 O_3 is not uncommon. In the Netherlands, O_3 concentration was high enough to reduce yields of potato, beans and poplars. In several locations in U.K., O_3 concentration exceeded 400 mg/m^3 in 1976.

Ozone also reacts with many fibers especially cotton, nylon and polyester, and dyes. The extent of damage appears to be affected by light and humidity. O_3 hardens rubber. A concentration of 0.3 ppm causes nose and throat irritation, 1-3 ppm causes extreme fatigue while a concentration of 9.0 ppm causes severe pulmonary edema.

O_3 is intimately connected with the life-sustaining process. Any depletion of ozone would, therefore, have catastrophic effects on life systems of the earth. Over the last few years, it could be realized that the O_3 concentration of earth's atmosphere is thinning out. What has caused this depletion?

In stratosphere, ozone absorbs UV rays causing the temperature inversion. This temperature inversion limits the vertical mixing of pollutants. However, in spite of this slow vertical mixing, some pollutants enter the stratosphere and remain there for years until they react with ozone and converted to other products. These pollutants thus deplete ozone in the stratosphere. Major pollutants responsible for this depletion are chlorofluorocarbons (CFCs), nitrogen oxides (coming from fertilizers) and hydrocarbons. CFCs are widely used as coolants in air conditioners and refrigerators, cleaning solvents, aerosol propellants and in foam insulation. CFC is also used in fire extinguishing equipment. They escape as aerosol in the stratosphere. Jet engines, motor vehicle, nitrogen fertilizers and other industrial activities are responsible for emission of CFCs, NO_x etc. The supersonics air crafts flying at stratosphere heights cause major disturbances in O_3 levels. The threat to ozone is mainly from CFCs, which are known to deplete O_3 by 14% at the current emission rate. On the other hand NO_x would reduce O_3 by 3.5%. The nitrogen fertilizers release nitrous oxide during de-nitrification. Depletion of O_3 would lead to serious temperature changes on the earth and consequent damage to life support system.

Depletion of ozone in stratosphere causes direct as well as indirect harmful effects. Since the temperature rise in stratosphere is due to heat absorption by ozone, the reduction in ozone would lead to temperature changes and rainfall failures on earth. Moreover one per cent reduction in O_3 increase UV radiation on earth by 2%. A series of harmful effects are caused by an increase in UV radiation. Cancer is the best-established threat to man. When the O_3 layer becomes thinner or has holes, it causes cancers, especially relating to skin like melanoma. A 10% decrease in stratospheric ozone appears likely to lead a 20-30% increase in skin cancer. The other disorders are cataracts, destruction of aquatic life and vegetation and loss of immunity. Nearly 6,000 people die of such cancers in USA each year. A 7% increase in such case has been reported from Australia and New Zealand.

Protecting ozone layer: Scientists throughout the world have shown a great concern over the thinning of ozone layer. This has led to a series of conferences on this issue. The first global conference on the depletion of ozone layer was held in Vienna (Austria) in 1985, the year, scientists discovered hole in South Pole. British team discovered a hole in ozone layer as large as that of the United States. This was followed by Montreal Protocol in 1987, which called for a 50% cut in the use of CFCs by 1998 reducing it to the level of 1986. Many countries including India did not sign the Protocol. India did not see any rationale as its release of CFCs is just 6,000 tones a year, equivalent to one and a half day's of world total. In our country per capita consumption of CFC is 0.02 kg. against 1 Kg. of developed world. CFCs are mainly the problem of developed world, as 95% of CFCs is released by European counties, U. S. A, USSR and Japan. USA alone releases 37% CFCs (producing CFC worth 2 billion dollar). The three-day international conference on Saving the ozone layer was organized jointly in London in March 1989 by the British Government and the UNEP. This conference highlighted the global problem created by the developed world, which in turn, is trying to dictate its terms to the developing countries for CFC pollution. The Montreal Protocol was initially signed by 31 countries.

There was held another international conference on ozone at Helsinki in May 1989 to revise the Montreal protocol. As many as 80 nations agreed to have a total ban on chemicals that cause ozone depletion by 2000 A.D. However, the conference backed away from a plan put forward by Dr. Mostafa Tolba, Executive Director, UNEP to set up an international climate fund. While the developing countries preferred to have the fund, the developed ones, including Japan, USA and UK rejected the plan. The life time of ozone is typically a few days to two months, and its global distribution is highly variable in time and space. The best O_3 data sets are from the last 20 years or so, during which time O_3 measurements were integrated into programmes at sites in the Global Atmosphere Watch (GAW). The world data center for surface ozone has recently become operational (1998).

HYDROCARBONS

Automobile exhaust contains a variety of hydrocarbons. These are emitted by evaporation of gasoline through carburetors, crankcase etc. The decomposition of organic waste and garbage also produces hydrocarbons like methane. From amongst many hydrocarbons, the chief air pollutants are benzene, benzpyrene and methane. In India, two and three wheelers are the main contributors, and in Delhi and Bangalore, emission form these accounts for about 65% of the total hydrocarbons. If unchecked, this may go up to 80% of the total hydrocarbons of air. About 40% of the vehicular exhaust hydrocarbons are unburned fuel components, the rest are the product of combustion.

They have carcinogenic effects on lung. They combine with NO_x under UV-component of light to form other pollutants like PAN and Ozone (photochemical smog) which cause irritation of eye, nose and throat, and respiratory distress.

Benzene a liquid pollutant is emitted from gasoline. It causes lung cancer. Benzpyrene is most potent cancer inducing hydrocarbon pollutant. It is also present in small amounts in smoke, tobacco, charcoal boiled stakes, and gasoline exhaust. Methane (marsh gas) is a gaseous pollutant, in minute quantity in air, about 0.0002% by volume. In nature this is produced during decay of garbage, aquatic vegetation etc. This is also released due to burning of natural gas and from factories. Higher concentrations may cause explosion. The excess of water seepage in filled up well and pits may lead to excess production of methane, which bursts with high sound and may cause local destruction. At high levels in absence of oxygen, methane may be narcotic on man.

FLUOROCARBONS

Fluorocarbons, in minute quantities are beneficial, helping prevention of tooth-decay in man but only in very minute amounts. However, higher levels become toxic, In India, there is a problem of fluorosis, as also in other countries as USA, Italy, Holland, France, Germany, Spain, Switzerland, China, Japan and some African and Latin American countries. In our country, it is a public health problem in states of Gujarat, Rajasthan, Punjab, Haryana, Uttar Pradesh, Andhra Pradesh, Tamil Nadu, Karnataka and some areas of Delhi. Fluorides in atmosphere come from industrial processes of phosphate fertilizers, ceramics, aluminum, fluorinated hydrocarbons (refrigerants, aerosol propellants etc.), fluorinated plastic, uranium and other metal. The pollutant is in gaseous or particulate state.

In particulate form it is deposited near the vicinity of emission, whereas in gaseous form it disperses over large areas. On an average fluoride level of air is 0.05 mg/m^3 of air. Higher values may also reach as in some Italian factories as much as 15.14 mg/m^3 of air. Resident of this area inhales about 0.3 mg/m^3 of fluoride daily. In air, fluoride chiefly comes from smoke of industries, volcanic eruptions and insecticide sprays. Fluorides enter leaves of conifers. Fluoride pollution in man and animals is mainly through water.

METALS

The common metals which cause air pollution are mercury, lead, zinc and cadmium. They are released from industries and by human activities in the atmosphere.

Mercury, a liquid volatile metal is present in air as a result of human activities as the use of mercury compounds in production of fungicides, paints, cosmetics, paper pulp etc. It has been reported that inhalation of 1 mg Hg/m^3 of air for three months may lead to death. Nervous system, liver, eyes are damaged. Infants may be deformed. Other symptoms of mercury toxicity, are headache, fatigue, anxiety, lethargy, loss of appetite etc.

Lead is added to gasoline to reduce knocking, and is emitted into the air with the exhausts as volatile lead halides (bromides and chlorides). About 75% of lead burnt in gasoline comes out as lead halides through tail pipe in exhaust gases. Of this about 40% settles immediately on the ground and the rest (60%) gets into air. The WHO prescribed level (2 mg/m^3) in air is already crossed in many countries of world. Lead inhalation causes reduced hemoglobin formation, thus leading to anemia. Lead compounds also damage RBC_s resulting in infection of liver and kidney in man. In automobiles lead accumulation increase emission of hydrocarbons.

Zinc, not a natural component of air, occurs around zinc smelters and scrap zinc refineries. Copper, lead and steel refineries also release some zinc in the air. Zinc in air occurs mostly as white zinc oxide fumes and is toxic to man.

Cadmium occurs in air due to industrial and other human activities. Industries engaged in extraction, refining, electroplating and welding of cadmium containing materials, and those in refining of copper, lead and zinc are the major source of cadmium in air. Production of some pesticides and phosphate fertilizers also emits cadmium to air. This metal is emitted as vapor, and in this state it quickly reacts to form oxide, sulphate or chloride compounds. Cadmium is poisonous at very low levels and is known to accumulate in human liver and kidney. It causes hypertension, emphysema and kidney damages. It may turn to be carcinogenic in mammals.

DEVELOPMENT OF INDIAN AIR QUALITY INDEX (IND - AQI)

Air is the prime resource for sustenance of life. With the technological advancements, vast amount of data about ambient air quality is generated to know the quality of air environment and to administer the appropriate corrective actions wherever necessary. Such an endeavor would result in encyclopedic volumes of data which may neither give a clear picture to a decision maker nor to a common man who simply wants to know how good or bad the air is. As for the general public, the questioner usually will not be satisfied with raw data, time series plots, statistical analysis and other complex findings pertaining to air quality. To address the above concerns, IIT Kanpur through a sponsored project from the Central Pollution Control Board, Delhi has proposed the Indian Air Quality Index (IND - AQI) in a simple and lucid terms.

A segmented linear function is used relating the actual air pollution concentrations (of each pollutant) to a normalized number. The basis for these linear functions (for this study) is arrived after considering such functions adopted by other countries (USEPA, 1976 and 1998; UK, 1998; Malaysia, 1997; GVAQI, 1997; Ontario, 1991; ORAQI, 1970). The pollutants included for the proposed IND - AQI are SPM, SO_2, NO_2, PM_{10}, CO and O_3. The following table presents the summary of the break point concentrations and AQI values for India (proposed) for all pollutants.

Table: Indian Air Quality Index for Different Pollutants

S.No.	Index	Category	SO_2 (24 hr avg) (mg/ m^3)	NO_2 (1-hr avg) (mg/ m^3)	SPM (24-hr avg.) (mg/ m^3)	CO (1-hr avg.) (mg/ m^3)	CO (8-hr avg) (mg/ m^3)	O_3 (8-hr avg.) (mg/ m^3)	PM_{10} (24-hr avg.) (mg/ m^3)
1.	0-100	Good	0-80	0-80	0-200	0-4	0-2	0-157	0-100
2.	101-200	Moderate	81-367	81-180	201-260	4.1-25	2.1-12	158-196	101-150
3.	201-300	Poor	368-786	181-564	261-400	25.1-35	12.1-17	197-235	151-350
4.	301-400	Very poor	787-1572	565-1272	401-800	35.1-75	17.1-35	236-784 (1-hr avg.)	351-420
5.	401-500	Severe	>1572	>1272	>800	>75	>35	>784 (1-hr avg.)	>420

Since all the pollutants are not measured under the National Ambient Air Quality Monitoring Program, it is necessary that at least three pollutant concentrations must be available to calculate and report the index. The over all AQI is based on maximum operator system where the maximum value of sub-index becomes the AQI. To reflect the attainment of NAAQS, the AQI is referred to as Good between the range 0 - 100. For the second break point (at the standard of USEPA), the AQI takes the value of 200 and referred to as Moderate. In absence of any other pollutant of any other pollutant health criteria in India, rest of categorization of Index is based on the USEPA Federal Episode criteria and Significant Harm Level.

IND-AQI is primarily a health related index with the following descriptor words: "Good (0 - 100)", "Moderate (101 - 200)", "Poor (201 - 300)", "Very poor (301 - 400)", "Severe (401 - 500)".

Although this website does not forecast the AQI for coming days, it however, does include the possible health effects so that the people can tke precautions; as in all likelihood the AQI is not going to change significantly on next day.

FEATURES OF WEB SITE

A website is developed for display of nation-wide air quality index. The display system is implemented using HTML, JavaScript, Java and CGI (Common Gateway Interface)/Perl script. As the air quality data become available, online calculation of AQI is carried out and displayed as an Air Quality Meter showing index value (with a pointer) with animation on the screen. The website is comprehensively designed to indicate the pollutant responsible for index and the pollutants exceeding the standards. The developed website facilitates for inter/intra city comparison of AQI using multiple windows for online as well as historical data of air quality. The general public can access the information through internet and other media agencies like newspaper, TV, radio can also download the information and disseminate the information.

CHAPTER 3

Energy from Biomass / Waste

As we all know that the sun is the largest source of energy. However, energy from wind, flowing water, geothermal sources and from biomass are other perpetual and renewable sources of energy worldwide. Energy is the main source of economic growth of a nation and reflects the state of development of a State. However, our conventional sources of energy are gradually decreasing and our dependence on them is greatly increasing day by day. A study says that our requirements of energy are doubling in every 15 years. Search for the alternative sources of energy is on and scientists say that hydrogen gas is the fuel of future. The most easy and best way to get more energy with the minimum environmental impacts is to cut its wastage. This can be achieved by reducing energy consumption and by improving the efficiency and doing more with less. Improving energy efficiency will reduce environmental damage as less of each energy resource would provide the same amount of the useful energy. Using it more efficiently would save money, provide more jobs and promote more economic growth. It is believed that only in USA such a large amount of energy is wasted which can be used by two third population of the world. In the countries like Japan, Sweden and some other industrialized European countries, some one third to two third less energy per person is used than the Americans. The sources which are used for obtaining energy differ in different countries. In United States about 92% of the known reserves and potentially available energy reserves are perpetual and renewable, the remaining is constituted by coal (5%), oil (2.5%) and Uranium (0.5%). In India about 40% of our energy requirements are obtained from firewood, agricultural waste, cowdung, animal power and other manual labour. The hydroelectric power and thermal power are the other important sources of energy. The hydroelectric power supplies about 20% of the world's electric energy and 6% of total commercial energy, however, Norway gets all its electricity from hydroelectric power, Switzerland gets about 74% and Austria about 67%

from hydroelectric power. The other sources of energy like tidal power, wind power and geothermal energy have their limitations and it may not compete economically with other alternatives. Biotecnology can play an important role in solving these problems and increase the acceptability of biomass and biogas as another viable energy alternatives for the future to come. It will not only be energy efficient but also be eco-friendly.

Biomass

Biomass is organic plant matter produced in the presence of solar energy by a process known as photosynthesis. It not only includes live plant matter, wood but also agricultural waste, manure and garbage. Some of it can be burned as solid fuel or converted into more convenient gaseous or liquid bio-fuel. Biomass, mostly from the burning of wood and manure to heat buildings and cook food supplies half of the energy used in the less developed countries and only 11 % towards the total world energy. Biomass is a renewable source of energy as long as trees and plants are not harvested faster than they grow back.

INTELECTUAL PROPERTY RIGHTS

Countries with innovative local industries almost invariably have laws to foster innovation by regulating the copying of inventions, identifying symbols, and creative expressions. These laws encompass four separate and distinct types of intangible property — namely, patents, trademarks, copyrights, and trade secrets, which collectively are referred to as "intellectual property."

Intellectual property shares many of the characteristics associated with real and personal property. For example, intellectual property is an asset, and as such it can be bought, sold, licensed, exchanged, or gratuitously given away like any other form of property. Further, the intellectual property owner has the right to prevent the unauthorized use or sale of the property. The most noticeable difference between intellectual property and other forms of property, however, is that intellectual property is intangible, that is, it cannot be defined or identified by its own physical parameters. It must be expressed in some discernible way to be protect able.

All four types of intellectual property are protected on a national basis. Thus, the scope of protection and the requirements for obtaining protection will vary from country to country. There are, however, similarities between national legal arrangements. Moreover, the current worldwide trend is toward harmonizing the national laws.

Intellectual Property Rights in India

India showed signs of resistance to quick enforcement of international intellectual property right (IPR) protection laws as demanded by the developed countries, particularly the US. But, India was required to implement WTO-standard IPR protection laws by 2005. It must be acknowledged that there has been remarkable progress in IPR protection in the field of software and cinema products.

India's general argument was that it does acknowledge in principle the case for strict IPR protection, but this can be done only in phases suited by its own ground reality. The reality is that absence of international IPR protection for some decades has spawned employment for millions, so an overnight clampdown on IPR violators would foment social unrest.

However, under pressure from its own domestic industry and the United States, India strengthened its copyright law in May 1994, placing it at par with international practice. The new law, which entered into force in May, 1995, fully reflects the provisions of the Berne Convention on copyrights, to which India is a party. Based on its improved copyright protection, India's designation as a "priority foreign country" under the United States' Special301 list was revoked and India was placed on the "priority watch list." Copyright enforcement is also rapidly improving.

Classification of copyright infringements as "cognizable offenses" expands police search and seizure authority. While the formation of appellate boards under the new legislation should speed prosecution, local attorneys indicate that some technical flaws in the laws, which require administrative approval prior to police action, need to be corrected.

Trademark protection is considered good by the US authorities, and could be raised to international standards with the passage of a new trademark bill that codifies existing court decisions on the use and protection of foreign trademarks, including service marks.

Enforcement of trademark owner rights had been weak in the past, but is steadily improving as the courts and police respond to domestic concerns about the high cost of piracy to Indian rights' holders.

India's patent protection is weak and has especially adverse effects on international pharmaceutical and chemical firms. Estimated annual losses to the US pharmaceutical industry due to piracy are $450 million, but Indian authorities have a different point of view.

India's patent act prohibits product patents for any invention intended for use or capable of being used as a food, medicine, or drug or relating to substances prepared or produced by chemical processes. Consequently, many drugs invented by foreign companies are widely reproduced.

Processes for making drugs are patentable, but the patent term is limited to the shorter of five years from the grant of patent or seven years from the filing date of the patent application.

Product patents in other areas are granted for 14 years from the date of filing. However, as a signatory to the Uruguay Round of GATT, including its provisions on TradeRelated Intellectual Property Rights (TRIPS), India must introduce a comprehensive system of product patents.

Patents: Society's Contract with Inventors

Patent systems encourage the disclosure of information to the public by rewarding an inventor for his or her endeavors.

Although the word "patent" finds its origins from documents issued by the sovereign of England in the Middle Ages for granting a privilege, today the word is linked synonymously with this exclusive right granted to inventors. The World Trade Organization (WTO) Agreement on Trade-Related Aspects of Intellectual Property Rights (TRIPS) provides the international standard for duration of patent exclusivity, which is 20 years from the date of filing. After the January 1, 2000, implementation date, all WTO members will be obligated to meet this standard. Under all patent systems, once this period has expired, people are free to use the invention as they wish. The benefits of an effective patent system can be partially illustrated as follows:

A patent rewards the investment of time, money, and effort associated with research. It stimulates further research as competitors invent alternatives to patented inventions, and it encourages innovation and investment in patented inventions by permitting companies to recover their research and development costs during the period of exclusive rights.

The limited term of a patent also furthers the public interest by encouraging quick commercialization of inventions, thereby making them available to the public sooner rather than later. Patents also allow for more latitude in the exchange of information between research groups, help avoid duplicative research, and, most importantly, increase the general pool of public knowledge.

Although the right conferred by a patent is a right to exclude others from making, using, or selling a patented invention during the term of the patent, it is important to understand that a patent does not necessarily give the patent owner the right to make, use, or sell the invention himself or herself. For example, the owner of a patent for an improved method of producing a chemical compound would not be free to sell the compound made using the patented method if the compound is itself patented by someone else.

Although all WTO members are subject to patent provisions in the TRIPS Agreement, patents are granted under national laws and, therefore, the rights are also national in scope. Thus, a U.S. patent can be asserted only against infringing conduct in the United States. In most countries these rights are enforceable by civil rather than criminal proceedings.

Accordingly, enforcement falls solely to the patent owner. In general, any act of making, using, or selling the patented invention without permission infringes that patent, whether it be by the state, a corporation, or an individual. Any such infringing act will give rise to liability, regardless of the infringer s intent or lack of knowledge of the patent. Remedies for patent infringement can include injunctions, orders to deliver up or destroy infringing articles, and compensation for damages suffered by the patentee or profits made by the infringer.

An issued patent remains open to attack for invalidity, and it is a common defense for an alleged infringer to assert that the patent is invalid. Typically, patents are challenged on the ground that the claimed invention was invented by someone other then the named inventor or that the invention would have been obvious to persons skilled in the relevant technology.

The limited term of a patent also furthers the public interest by encouraging quick commercialization of inventions, thereby making them available to the public sooner rather than later. Patents also allow for more latitude in the exchange of information between research groups, help avoid duplicative research, and, most importantly, increase the general pool of public knowledge.

Although the right conferred by a patent is a right to exclude others from making, using, or selling a patented invention during the term of the patent, it is important to understand that a patent does not necessarily give the patent owner the right to make, use, or sell the invention himself or herself. For example, the owner of a patent for an improved method of producing a chemical compound would not be free to sell the compound made using the patented method if the compound is itself patented by someone else.

Although all WTO members are subject to patent provisions in the TRIPS Agreement, patents are granted under national laws and, therefore, the rights are also national in scope. Thus, a U.S. patent can be asserted only against infringing conduct in the United States. In most countries these rights are enforceable by civil rather than criminal proceedings.

Accordingly, enforcement falls solely to the patent owner. In general, any act of making, using, or selling the patented invention without permission infringes that patent, whether it be by the state, a corporation, or an individual. Any such infringing act will give rise to liability, regardless of the infringer s intent or lack of knowledge of the patent. Remedies for patent infringement can include injunctions, orders to deliver up or destroy infringing articles, and compensation for damages suffered by the patentee or profits made by the infringer.

An issued patent remains open to attack for invalidity, and it is a common defense for an alleged infringer to assert that the patent is invalid. Typically, patents are challenged on the ground that the claimed invention was invented by someone other then the named inventor or that the invention would have been obvious to persons skilled in the relevant technology.

Article 27 of the TRIPS (Trade-Related Aspects of Intellectual Property Rights) Agreement provides that WTO member states shall provide patents for any invention, either a product or a process for creating a product, "provided that they are new, involve an inventive step, and are capable of industrial application." In other words, to be patentable, an invention must be novel, useful, and no obvious. A prerequisite to patentability is that the invention must be capable of some practical application. This emphasizes the importance the patent system puts on usefulness. Although this principle remains constant, the phraseology used within the legislation of particular countries varies; for example, in the United States, patentable subject matter must be "useful," whereas in the United Kingdom it must be capable of "industrial application."

The invention must be new — that is, the subject matter of the invention is not or cannot be inferred to be part of what is already known. This is commonly referred to as the "novelty" requirement. New or novel in this context means "new to the public." Therefore, something that has previously been used or known but has not been made available to the public (for instance, if it has been kept a secret) is not a bar to patentability.

The invention must also be no obvious. This prevents someone from taking advantage of the patent system and obtaining protection for something that is a mere extension or trivial variation of what is known. Generally the test for inventiveness, or "no obviousness," is based on what a reasonable person skilled in the field to which the invention pertains, at the time the invention was made, would consider to be no obvious.

The TRIPS Agreement provides a transitional period for developing economies that do not currently provide product patent protection in the areas of agro-chemicals or pharmaceuticals. In fact, most already do because of the development benefits to the biotechnology sector from full patent protection. Process patent protection does not encourage investment because of the difficulty of enforcing a process patent. It is particularly difficult to enforce a process patent because the burden of proof to show that the patent has been infringed is on the patent owner. The patent owner must prove that a particular manufacturing process (that is, the process covered by the patent) was used to manufacture the particular chemical. This can be very difficult to show where there are many possible process variants and where access to the potential infringer s facility is not available. In practice, this is done by looking for trace impurities that are characteristic of the manufacturing process. One can imagine how complex the issues can become if, for instance, a patent protects a pharmaceutical that is made in a country where there is no protection for pharmaceuticals and then is exported to a second country that provides protection only for manufacturing processes.

Over the past 15 years or so, many countries have changed from "process" to "product" patenting, and we expect all WTO members to upgrade their patent laws within the next few years because, under the WTO TRIPS Agreement, member states must provide full product patent protection no later than January 1, 2005.

Not only are the utilitarian aspects of new and useful inventions patentable, but many countries extend patent protection to novel, ornamental industrial designs. In the United States, this form of protection is known as a design patent, while in many European countries, the property right in an industrial design is referred to as a design model.

In addition to such usual subjects of patent protection as devices, chemical compositions, and processes, some countries provide patent protection for living matter. For example, asexually reproduced varieties of plants, excluding bacteria, uncultured plants, and tuber propagated plants, can be protected, as can sexually reproduced plants (by seed), excluding bacterial, fungi, and first-generation hybrids. The TRIPS Agreement does not require protection for new living matter or plant varieties, but WTO members may join the International Union for the Protection of New Varieties of Plants, or UPOV.

TRADEMARKS AND SERVICE MARKS: IDENTIFYING THE SOURCE

Trademarks and service marks are primarily intended to indicate the source of goods and services and to distinguish the trademarked goods and services from others. They also symbolize the quality of the goods or services with which they are used. Most trademarks and service marks (called "marks") are words, but they can be almost anything that distinguishes one product or service from another, such as symbols, logos, sounds, designs, or even distinctive nonfunctional product configurations.

The TRIPS Agreement extends the same level of recognition and protection for service marks as for trademarks (TRIPS Agreement Articles 15, 16). In some countries, registration of a mark may not be required to protect the mark, but in any case WTO members are obligated to provide protection for well-known trade or service marks. Because determinations of whether a mark is well known in the relevant sector of the public are made on a case-by-case basis, firms may find it desirable to register well- known marks. For marks that are not well known, countries may require the owner of the mark to register the mark with the national trademark office before protection in that country is granted.

The duration of protection afforded a mark varies greatly from country to country. Registrations are issued for finite periods of time. However, because of the fundamental purposes of marks namely, avoiding public confusion, encouraging competition, and protecting the owners' goodwill registrations may be renewed and thus extend indefinitely as long as the marks are used.

The owner of a mark may preclude others from using a similar mark if such use is likely to cause confusion in the minds of purchasers. Determining whether two marks are so similar as to be confusing usually involves a multi-factor analysis that compares the parties marks, their goods or services, their advertising and trade channels, the defendant s intent in choosing its mark, and the presence or absence of actual confusion.

A Spectrum of Protection for Trademarks and Service Marks

As in other intellectual property areas, trademark and service mark legislation is national in origin but must comply with provisions of the TRIPS Agreement.

Some countries grant rights to the first person to use the mark in the course of business, while other countries grant rights to the first person to obtain a registration in that country.

In "first-to-use" countries, rights may subsist without registering the mark with the national trademark office. However, registration is still desirable because it is presumptive evidence of the validity of the mark and the owner's right to use that mark. It also appears on the national register of marks, providing notice to the world of the owner's use and claim of ownership. Under TRIPS, actual use of a trademark shall not be a condition for filing an application for registration of a trademark or service mark. Following the use or registration of a trademark, the owner must use the mark or it may become subject to attack by others on the grounds that the owner has abandoned it.

At a minimum, most countries require that a mark be distinctive; that is, it should be capable of distinguishing the goods or services of the owner of the mark from the goods or services of others. A mark may include any original combination of numbers, letters or other symbols, colors, or musical tones. To determine whether a mark meets this test, one must determine the strength of the mark.

Originality As the Key to Copyright

To secure copyright protection, the work in question must be an original work of authorship fixed in a tangible medium of expression. Works of authorship that fall within this definition may include:

- Literary works (including computer programs);
- Musical works and accompanying lyrics;
- Dramatic works and dialogue;
- Pantomimes and choreographic works;
- Pictorial, graphic, and sculptural works;
- Motion pictures and other audiovisual works; and
- Sound recordings

It is important to note that the laws of many countries do not limit the type or form of work because authors are continuing to invent new ways of expressing themselves.

The test for the originality of a work is usually two- pronged. First, the work of authorship must originate from the author, in the sense that it must have actually been independently created by the author and not copied from other works. Second, the work must contain a sufficient amount of creativity so as to be more than trivial.

Trade Secrets: The Competitive Edge

A trade secret is information that is secret or not generally known in the relevant industry and that gives its owner an advantage over competitors. Trade secret protection exists as long as the information is kept secret or confidential by its owner and is not lawfully and independently obtained by others. Examples of trade secrets include formulas, patterns, methods, programs, techniques, processes, or compilations of information that provide one s business with a competitive advantage. The owner of a trade secret may recover damages resulting from the improper disclosure or use of its trade secret by another.

Costs vs. Benefits

As with all business-related activities, economics plays a large role in determining whether to protect intellectual property. Companies must weigh the potential value of an intellectual property right against both the probability of realizing that value and the costs of securing, enforcing, and maintaining that right.

There are no hard and fast rules that determine the potential value of a given intellectual property right. What is valuable to one individual or company may be worthless to another. There are certain obvious factors that contribute to the potential value of the intellectual property, including the potential value of exclusive or other rights, assignments, or licenses, cross-licenses, enforcement against infringers, and as collateral for securing financing.

Trademark or service mark may be a very valuable asset. For example, it is widely believed the German automobile manufacturer BMW purchased the British automobile manufacturer Rover primarily to obtain its portfolio of desirable trademarks including "Land Rover," "Range Rover," "Triumph," "Austin," and "MGB." On the other hand, a trademark may be virtually worthless if consumers associate it with poor quality.

4

CHAPTER

Environmental Ethics

The dictionary meaning of ethics is 'set of moral principles'. The word is derived from the *Greek* word ethos means customs and we know that it is beliefs, attitudes or norms that form the basis of customs. Our attitudes, we know are shaped to a considerable extent by social, economic and political factors. Environmental Ethics is a branch of philosophy concerned with the moral relation between human and the natural world. We need to understand each other, if no other reason than enlightened self-interest. But a nobler motive is the quest for an international environment ethic, which concerns itself with these global concerns: The humanity's relationship to the environment; its understanding and responsibility to nature and its obligation to leave some of the nature's resources for future generations. Environmental ethics have to sharpen the judgment of a person not to jeopardize the health and security of other fellow beings for the sake of material and political gains.

WHY TO FOLLOW ETHICS

The industrial revolution accelerated the development of economy in the west leading to industrial growth, and subsequently advocating concepts of profit and increase of share prices.

Ethics as an important issue in such considerations was not a matter of concern. The business for example, was not much concerned with ethical overtones. Profit maximization was the motive of the companies. However, during the last decade malpractices by the companies surfaced to the embracement of the stake holders. It became obvious that issue like transparency and accountability cannot be avoided; till then ethical consideration did not factor in business strategy.

Literature on ethics and business started appearing dealing with debate on compatibility of economic, social, environmental and ethical issue; debate sometimes becoming philosophical because ethics concern moral values are very much the domain of philosophy. Should ethics enter the realm of corporate behavior ? Should a corporation observe ethical code. In the Indian context we are not too worried about talking ethics. People do understand its necessity and importance. Swami vivekanada's saying to build character first and then learn whatever trade is very much relevant. It is taken from our ancient distinction between vidya which you get at the IITs and the Harvard Business School and the IIMs and so on and brahmvidya which you don't get there, which is knowledge of self consciousness, ethics, and that one to pick up".

THE BASIS OF ENVIRONMENTAL ETHICS

It is implicitly assumed that values of human health and environmental protection have more importance than our social behaviour. Environmental deterioration, ecological disasters and decay of spirit are inevitable, if we let scientific technology continue in its present course. Many societies in the past have committed themselves to a course that eventually led to their destruction (Like Roman civilization). Similarly, ours will experience catastrophe if it remains obsessed as it is today with the production of more power and more things.

What is needed today is to remind ourselves that nature can not be destroyed without mankind, ultimately being destroyed itself.

The emergence of environmental crisis prompted the rise of green movements. This was accompanied by the development of new environmental ethics, such as population control, biocentric ethics, ecocentric ethics and deep ecology. No doubt, unethical practices may give an initial advantage but it can tarnish company's image sooner or later. The companies must realize (some of them have realized now) that pressure groups are watching and have a voice because of their being voluntary character. Also the thinking at the top management has to be one with typical behavior.

The question is how to utilize the issue arising out of the ethic-environment debate as the basis for principles that have wide acceptance and than how to translate those principles into practice-are difficult task. Meanwhile, the companies have started showing awareness and few attempts towards observing ethical standard like (whatever the definition of these may be) we should preserve biodiversity; we should avoid or reduce the use of artificial fertilizers or pesticides.

This is a normative aspect of ethics because this is an appeal to follow certain norms. But why should we follow normative plains. In other words we have to go a step further to justify normative judgments in environmental matters like who decide the norms? As we know that today, the environmental damage, resource depletion, and host of associated problems pose a great challenge.

One view is that technology can be exploited to fix norms and also to mitigate environmental problems. But technology has limitations; many technological solution are not free from controversies like the issues of dams or genetically modified foods. Sometimes, (not always) scientific discoveries or technological solutions have resulted in a new set of problems. Technological experts or scientists then fall back to basic issues of values and ethics and other philosophical matters for guidance. But ethical and philosophical theories also have limitations in so far as the solutions of environmental problems are concerned. This is our dilemma. What we can do is to wait to learn more about the debate on technologies and ethics-environment relationship.

Social responsibility means, providing safety for the workers at workplace and concern for their health, reducing pollution and creating other welfare schemes in the areas like education and employment. A good social performance of an organization is a positive development in the evolutionary process of ethics. Also, government's regulations for better work conditions in an organization strengthened the issue of social responsibility. A company carrying out some sort of social responsibility and showing a concerns for public at large could gain a better image which would enhance the business prospects.

Upto this stage, the scholars of ethics believed that economic basis or economic ethics was still predominant. And the corporation still believed that being more socially responsible and being concerned for welfare of the society means spending more money resulting in lowering of profit.

Can a Manager Pursue Social Goals that Conflict with Economic Goals?

No doubt, there is always a temptation for the companies adopting short-term policies of profit maximization and not showing concerns for social and environmental issues that increase the cost. But today, in this era of information revolution, transparency can not be evaded. And the stake holders can not be taken for granted. Probably at this stage when the conflict between the economic and social goals was being debated, the issue of environmental values entered the arena.

It may or may not be coincidence that at this time the regulatory/ mandatory requirements became more stringent and the organizations now have had to incorporate environmental values in their governance. Thus it became clear that guide lines in respect of certain issues of social and environmental concerns are necessary for corporate function.

THE ETHICS OF POPULATION CONTROL

Ehrlich warned that the exponentially increasing population of the world would cause an increase in poverty, hunger and environmental deteriorations, as billions of people would devastate the remaining resources. It would be impossible for agriculture production to sustain growing population indefinitely. As he reasoned: A cancer is an uncontrolled multiplication of cells, the population explosion is an uncontrolled multiplication of people. Treating only the symptoms of cancer may make the victim more comfortable at first, but eventually he dies-often horribly. A similar fate awaits a world with a population explosion if only the symptoms are treated.

As possible solutions he suggested a change in tax laws, so that tax would increased instead of decrease with additional children, as well as national policy encouraging contraception, laws permitting abortion and distribution of information and birth control devices.

Exponential growth of population generates an exponential demand for food, supply of which depends on land, water, fertilizer and agricultural machinery, which depends in turn on capital growth and non-renewable resources, of which exits a finite stock. Thus, factors of growth are interrelated such that exponential growth as a whole becomes a threat to the survival of this planet. India accommodates about 16% of the world's population on just 2.45% of the world's land area with mere 1.5% of the world's income. This status is getting further deteriorated by still over increasing population. With present growth rate of 2.11%, we are adding 1.8 crore individuals every year and almost 35 individuals per minute.

LIFE BOAT ETHICS

Hardin (1974) in his famous "life boat ethics" refers to ethics that would be adopted when lifeboat space was insufficient to accommodate all passengers from a sinking ship a situation such as occurred when the titanic sank in the north Atlantic.

The rich nation of the world may be thought of as lifeboats with moderate numbers of rich persons on board, while the poor countries much more crowded lifeboats. The poor says Hardin continuously fall out of their lifeboats, swim for a while in the water outside and hoping to be admitted to a rich lifeboats. The ethical dilemma is that what should the passengers on a rich lifeboat do? Whether the passengers on the less crowded boat should help the swimmer or allow for admission to boat; like the Land of every nation, each lifeboat has only a limited carrying capacity.

If one follows Christian or Marxist ethics, we are bound to admit everyone to the boat since the needs of those inside and outside the boat are the same. Doing so, the boat is swamped and everyone drowns "complete justice complete catastrophe".

Hardin's proposed solution is to protect the survival of those on board, preserve the "safety factor" and admits no more people to the rich boats. He justifies this solution by explaining that if rich nations gave resources to poor ones or admitted many of their people through generous immigration policies, then disaster would eventually strike the countries. The prolific passengers would swamp the life boat. Hardin's words sound cold and inhuman, for they show no pity and recommend no help to the starving million at our door.

But in the view that it is more ethical to save a small population for a long time more in humane than the attempt to help everyone, only to have total disaster shortly thereafter.

BIOCENTRIC ETHICS : HUMAN CENTRED ETHICS

Biocentric ethics means life-centered ethics. Schweitzer (1923) describes the theory of "Reverence for life" that all of life is sacred and that we must live accordingly , treating each living being as an inherently valuable "will to live". Every living thing (every will to live) in nature is endowed with something sacred or intrinsically valuable and should be respected as such.

Taylor (1981) develops Schweitzer's life-centered system of environment ethics. He argues that each living individual is a "technological center of life" which pursues its own good in its own way and possesses equal inherent worth. Human beings are no more intrinsically valuable than any other living thing but should see themselves as equal members of Earth's community. For Taylor, all living beings from amoeba to humans are of equal inherent value.

Each living individual has a goal, and to have a goal implies a will or desire to attain it. One's goal is one's good, so all living things are inherently good. The chief proponents of this view usually tie their doctrine to re-incarnation, and transmigration of souls.

ECO CENTRISM: THE LAND ETHIC

In "Land Ethic" Leopold (1949) argued that we must begin to realize our symbiotic relationship to Earth, so that we value "The Land" or biotic community for its own sake. We must come to see ourselves, not as conquerors of the land but rather, as plain members and citizens of the biotic community. It implies respect for his fellow-members and also respect for the community as such.

Leopold urged a new standard for ethics. In judging the very meaning of "right" and "wrong", he said, we should put the living land at the center *"A thing is right when it tends to preserve the integrity, stability and health of biotic community. It is wrong when it tends otherwise".*

In this regard, Leopold offers a holistic biocentric ethic, in contrast to the mainly atomistic, anthropocentric, ethics familiar in all the western traditions. It is extremely important to develop an alternative view point because exclusive attention to what seems to be good for humans in the short term has proven ruinous and promises to inflict even worse environment damage in the future. To lead us out of this anthropocentric morass, it might seem that a land ethics of holistic biocentrism could be an important guide.

DEEP ECOLOGY

Deep ecology movement pioneered by Norweign philosopher Arne Naess in 1973 in his article "The Shallow and the Deep: Long Range Ecology Movement" and elaborated by Bill Devall and George Sessions in 1984. The movement advocates a "biocentric equality" that is- the notion that human are part of nature, and as such, have no more right to exploit other species as those other species have to exploit humans.

In this respect, deep ecology is similar to Asian traditions of thought such as Buddhism, and Hinduism whose fundamental teaching is that humans are part of nature, not above it: humans have no right to kill other creatures, except as it become necessary to save a life.

Ecology holds that all of us-humans, nonhumans, and biotic communities are intrinsically related to one another. Naess contrasts his deep ecology movement with the dominant world view of technocratic-industrial society which regard humans as isolated and fundamentally separate from the rest of nature, as superior to, and in-charge of, the rest of creation. But the view of humans as separate and the superior to the rest of Nature is only part of larger cultural patterns. with dominance of humans over non-human Nature, masculine over the feminine, wealthy and powerful over the poor. Deep ecological consciousness allows us to see through these erroneous and dangerous illusions.Deep ecology call for the promotion and greater protection of biodiversity and reduction in human population.

It's motto is Simple in needs and Rich in ends.

WHAT WE CONCLUDE FROM THE ABOVE

1. Our individual life, institutional activities, formal and informal education all need to be more environmental based and environmental oriented.
2. The changing relationship between man and nature is not merely an ecological phenomena. It has profound economic, social and cultural consequences.
3. The sacred quality of nature must be given back to it.
4. Today there is a crisis of perception everywhere. We are living in a globally connected world and any action for future demands an ecological attitude, assuming evolutionary changes and continuous creative activity on the part of man.
5. We have to evolve a a more balanced way of thinking, feeling and acting towards the environment.

   ACCEPTING "GREEN LIFE STYLE" IS THE NECESSACITY OF OUR TIME

5 CHAPTER
Debt and Environment

At first glance, it may seem like separate issues, but environment issues and poverty/debt are very much related. In fact, as Juilee 2000 explains, it affects all of us. Basically, the more the developing countries stay in debt, the more they will feel that they need to milk the earth resources for the hard cash they can bring in, and also cut back on social, health, environmental conservation, employment and other important programs. In fact, even the World Wild life Fund has started a campaign for the poor, acknowledging the link between poverty and environmental resource problems.

Brazil's IMF debt and financial problems have several affected a project to save the Amazon rainforest. And then there are many situations where the poor often have indigenous and traditional knowledge of their environment and are the best maintainers of it. However, when poverty has been imposed on them and international trade agreements force them to abandon their ways, much is lost.

Indian activist and scientist, Vandana Shiva, for example, shows in her book *Stolen Harvests* (South End Press, 2000) that those who are often beneficial to the environment have been forced into poverty due to politics and economics such as concentrating land rights, pressure from industry to use the environment for other purposes or in other ways, etc. Industrialization is often to blame for losses in diversity. Excessive debt burden means that it becomes harder to sustain the environment. Expensive aid and development programs from Europe have been found to be destroying parts of the environment in developing countries and affecting local and indigenous people into further poverty and misery. A lot of this is due to lack of consideration and communication with the people who are directly in the line of these development programs. According to a Christian Aid report, industrialized nations should be owing over 600 billion dollars to the developing nations — three times as much as the conventional debt that developing countries owe the developed ones.

In the Kyoto Conference On Climate Change, USA complained about the unfairness because developing nations did not have to reduce emissions like the developing nations were to do. Well, the report above, makes the point that many developing countries have also been trying to make; that the environmental consequences of the policies of industrialized nations have had a large, detrimental and costly effect on developing countries — especially the poor in those countries, that are already burdened with debt.

DEBT RELIEF AND NATURAL DISASTERS

When poor countries face natural disasters, such as hurricanes, floods and fires, the cost of rebuilding becomes even more of an issue when they are already burderned with debt. Often poor countries have had to suffer with many lost lives and some aid while still paying millions a week back in the form of debt repayment.

DEBT RELIEF AND THE FLOODS OF MOZAMBIQUE AND MADAGASCAR

The worst floods for 50 years and a devastating cyclone in the first three weeks of February, 2000 in Mozambique has been met with a slow response by the international community to provide much needed assistance and aid. As many as 300,000 were feared to have lost everything, while there werefears of more floods The floods are also thought to have moved landmines from previously known minefield areas to areas that have already been cleared. Mozambique is one of the world's most heavily mined countries.

As with the effects from Hurricane Mitch, debt burdens once again have come to the fore. Even Mozambique's government has been urging a cancellation of the country's debts to help use those saved resources for rebuilding the destroyed infrastructure.

The UN also estimates that 600,000 people have been affected in Medagaskar, the world's fourth largest island. That is twice as many in Mozambique. Two cyclones tore through the island. One of them made its way on to Mozambique resulting in the damage that filled headlines. However, a second cyclone that hit Madagascar didn't get to Mozambique.

DEBT RELIEF AND HURRICANE MITCH

November 1998, the Central America witnessed a deadliest storm which has been terrible. A UN report estimates that the destruction caused will set back development in this region by 20 years. The cost of rebuilding after Hurricane Mitch has highlighted the problems of debt repayment and debt relief.These countries are still facing a repayment of about $ 200 million a day from Honduras and Nicaragua, two of the worst hit areas).

In May 2001, the international community also agreed "to seek a moratorium on debt service payments for the world's most highly — indebted countries in "exceptional" situations — such as those plagued by civil wars, floods and natural disasters — and to facilitate access to debt relief for post-conflict countries." Hopefully then, in cases like the above, there might be a chance for some rest bite.

CAUSES OF THE DEBT CRISIS

Third world debt has long been recognized as a major obstacle to human development. Many other problems have arisen because of the enormous debt that third world countries owe to rich countries. Debt has impeded sustainable human development, security and political or economic stability. How has this happened?

A CONTINUING LEGACY OF COLONIALISM

The historic causes of third world debt is introduced in a working paper from the development organization, the *South Centre*. It summarizes how the developing countries' debt is partly the result of the unjust transfer to them of the debts of the colonizing States:

The history of third world debt is the history of a massive siphoning-off by international finance of the resources of the most deprived peoples. This process is designed to perpetuate itself, thanks to a diabolical mechanism whereby debt replicates itself on an ever greater scale, a cycle that can be broken only by cancelling the debt.

According to a new Working Paper on "Effects of debt on human rights" prepared by Mr. El Hadji Guissé for current UN Sub Commision on Human Rights (E/CN.4/Sub.2/2004/27), the developing countries' debt is partly the result of the unjust transfer to them of the debts of the colonizing States!

A sum of US$ 59 billion external in public debt was imposed on the newly independent States in 1960. With the additional strain of an interest rate unilaterally set at 14 per cent, this debt increased rapidly. Before they had even had time to organize their economies and get them up and running, the new debtors were already saddled with a heavy burden of debt.

ODIOUS DEBT

Odious debt is unfair debt resulting from illegitimate loans. A useful summary from *Jubilee USA*: Odious debt is an established legal principle. Legally, odious debt is debt that resulted from loans to an illegitimate or dictatorial government that used the money to oppress the people or for personal purposes. Moreover, in cases where borrowed money was used in ways contrary to the people's interest, with the knowledge of the creditors, the creditors may be said to have committed a hostile act against the people. They cannot legitimately expect repayment of such debts.

Many poor countries today have started their independent status with heavy debt burdens imposed by the former colonial occupiers. South Africa as another example, has found it now has to pay for its own past repression: the debts incurred during the apartheid era are now to be repaid by the new South Africa.

But it is not just South Africa paying for this; surrounding countries that have been destabilized from this are paying debts incurred to deal with it. The organization Action for Southern Africa summarizes this clearly, albiet in a report from 1998.

This report estimates "apartheid-caused debt" at £28 billion [about $46 billion at the time the report was written]. That is the £11 billion [$18 billion] that South Africa borrowed to maintain apartheid, and the £17 billion [$28 billion] that the neighbouring states borrowed because of apartheid destabilisation and aggression. This is 74% of the present regional debt of £38 billion [$62.5 billion].

Apartheid wrought vast destruction across the region; now the people of Southern Africa want to rebuild. In a remarkable spirit of reconciliation, the people of Southern Africa want to forgive the horrors of the past and look forward. But the banks, international financial institutions, and individual countries which lent to both sides in the apartheid war are demanding repayment.

After the Second World War, the United States allowed Britain to repay debt at a very low rate so that it could rebuild. In 1953, the victorious allies met in London to cancel most of Germany's debt, so that it could rebuild.

Now the nations of Southern Africa want to rebuild a post-apartheid society, but the creditors of today, are not willing to offer them the space Britain received from the US and the Allies gave to Germany. Instead they are demanding that the states of Southern Africa pay three to five times the level that Britain or Germany paid after World War II.

The report also adds that countries further away, such as Tanzania, also felt the effects and had invested substantial sums (about $800 million for Tanzania) to appose apartheid.

Various other nations have found that they have to pay debts incurred by their previous military dictators (many of which were installed as clients of the rich countries.

MISMANAGED LENDING

Further debt resulted from mismanaged spending and lending by the West in the 1960s and 70s. 1960s saw the US spend more than it had, resulting in the printing of more dollars.

Oil-producing countries, pegged to the dollar were affected as the value of the dollar decreased. In 1973, the oil-producing countries hiked their prices as a result, earning a lot of money, which they put in to western banks.

Interest rates started to plummet resulting in more lending by banks to try and prevent a crisis. A lot of the borrowed money went to western-backed dictators, resulting in little benefit for most people.

In 1982 Mexico defaulted on its debt payment, threatening the international credit system.

The IMF and World Bank stepped in to Mexico and other nations facing similar problems, prescribing their loans and structural adjustment policies to ensure debt repayment. The poor have suffered the most as a result of the harsh conditions of structural adjustment. Poor countries have soft currencies (values which can fluctuate). Debt crises can also occur just by the value of the developing country's money going down, which can be due to a variety of other inter-related factors. Paying off loans implies earning foreign exchange in hard currencies. Combined with falling export prices for many poor countries, debts become even harder to pay off.

Refinancing loans implies taking on new debts to service the old ones. Structural adjustment advice in the past from the IMF and others, has led to the cut back on important spending such as health, education, in order to help repay loans. This has implied a downward spiral and further poverty. Most loans to the third world have to be paid back in hard currencies (which do not usually change too much in value, *e.g.* the Japanese Yen, the American Dollar, etc.).

Economists often refer to a *moral hazard* of forgiving debts, because it may encourage people to take on new loans and refuse to pay. Yet, as *Action for Southern Africa* also noted in the above-mentioned report about Southern Africa's odious debt, the problem is not necessarily with borrowers, but with lenders:

To repay odious debts is to encourage lending to pariah regimes. If banks could lend to apartheid South Africa in the face of global opposition and global calls for sanctions, and still collect on the loans, then the signal to international banks is that they can lend to any regime, no matter how repugnant. There is a "moral hazard" here: that we will encourage immoral lending.

THE WORLD'S POOR ARE SUBSIDIZING THE RICH

Another cause for large scale debt has been the corruption and embezzlement of money by the elite in developing countries. These moneys are often placed in foreign banks (and used to loan back to the developing countries). Many loans also come with conditions, that include preferential exports etc. In effect then, more money comes out of the developing countries than is given in. This depresses wages even further due to the spiraling circle downwards to ensure that enough exports are produced. (See the structural adjustment section on this web site for more on that aspect.)

J.W. Smith, from the Institute for Economic Democracy, is worth quoting at length:

Susan George, in her 1992 book, *Debt Boomerang: How Third World Debt Harms Us All*, calculated a net of $418 billion borrowed funds flowed right back north between 1982 and 1990.

The world's poor are subsidizing the rich. The net gain to the over-capitalized countries (loss to the under-capitalized ones) of $418 billion between 1982 and 1990 is more than double what was spent to rebuild Europe after World War II. "Capital flight from Mexico between 1979 and 1983 alone [was] $90 billion — an amount greater than the entire Mexican debt at that time."

The world's powerless cannot obtain their share of capital, high paying jobs, and markets. Thus, they trade their valuable resources for products manufactured by well-paid labor in the over-capitalized countries. Just as cheap imported agricultural products destroy an undeveloped country's agricultural economy, imported consumer goods forestall the building of industry to produce these products regionally and build an internal market economy.

If a loan is to be of lasting value to the country to which it is granted, it must be put to *productive,* not unnecessary consumptive, or wasteful use. Only by building the tools of production (industry) instead of spending borrowed funds on consumption can a society become self-sufficient, build an internal market economy, gain equality in world trade, and eliminate poverty. *J.W. Smith, The World's Wasted Wealth 2, (Institute for Economic Democracy, 1994), pp. 139-141.*

BACKBONE TO GLOBALIZATION

The economic decisions and influence in various international agreements, treaties and institutions by the wealthy and powerful nations also help form the backbone of today's globalization. That such immense wealth and prosperity for some have come at a time when most nations in the world have steeped into further poverty and debt is no coincidence.

The policies of those who have the power and influence have been successful to help raise standards for some in their own nations, but at a terrible cost. Rich nations as well as poor incur debts, but often the wealthier and more powerful ones are able to use various means to avoid getting into the dilemmas and problems the poor nations get into. In fact, the following summarizes it quite well using the U.S. as an example:

THE GENERAL AGREEMENT ON TARIFFS AND TRADE (GATT)

The General Agreement on Tariffs and Trade (GATT) was first signed in 1947. The agreement was designed to provide an international forum that encouraged free trade between member states by regulating and reducing tariffs on traded goods and by providing a common mechanism for resolving trade disputes. GATT membership now includes more than 110 countries.

Consideration of GATT's relationship to environmental policy is an emerging concern in trade and environmental policy circles. Until the recently concluded Uruguay Round of GATT negotiations (1986-1994), the word environment did not appear in the GATT text. Several provisions and sections of GATT may be relevant to environmental issues, however. The following sections of GATT are often referenced in the examination of trade-environment issues:

Article I General Most-Favored-Nation Treatment;

Article III National Treatment on Internal Taxation and Regulation;

Article XI General Elimination of Quantitative Restrictions;

Article XIII Non-discriminatory Administration of Quantitative Restrictions;

Article XVI Subsidies;

Article XX General Exceptions;

The GATT Final Act Embodying the Results of the Uruguay Round contains several other relevant items which include:

- the Trade-Related Aspects of Intellectual Property Rights;
- an Agreement on Subsidies and Countervailing Measures that permits some environmental subsidies in section 8.2;
- and the Agreement Establishing the Multilateral Trade Organization.

Explicit environmental concern related to GATT has grown in the last few years. Although formally organized in 1971, a GATT committee on environment met for the first time in 1991. Also in 1991 a GATT dispute-resolution panel made a ruling in the tuna-dolphin case, a dispute between the United States and Mexico over limiting imports based on fishing practices. In 1992, GATT produced "Trade and the Environment," a report that offered GATT's recommendations on the relationship between international trade and environmental measures. The recent cases, the 1992 report, and growing international consideration of the relationship between trade and environmental policy have focused attention on GATT's influence on international environmental agreements. In some ways, GATT is seen as potentially limiting or barring trade provisions in environmental agreements. Environmental, legal, and other experts have also called for reform of GATT to accommodate international concern for environmental issues.

TRADE-ENVIRONMENT TENSIONS

Under GATT rules, countries are supposed to treat imported products "no less favorably" than comparable goods produced domestically. However, developing countries are concerned that some ostensibly environmental standards may actually be "green" protectionism—standards designed to favor domestic producers over foreign competitors.

The Draft Final Act of the Uruguay Round of the GATT contains new rules on the use of technical regulations and standards. These new rules would cover, for example, environment, health, and safety regulations pertaining to agricultural products, chemicals, and motor vehicles. The rules would require countries to use the least trade-restrictive means for achieving environmental objectives and, in most cases, to use relevant international standards instead of national ones.

Environmentalists fear that international standards will become a ceiling rather than a floor, and will exert downward pressure on the environmental standards of developed countries. They are also concerned that a least trade-restrictive rule could be interpreted narrowly by GATT dispute panels, threatening domestic regulatory regimes. Both developed and developing countries are concerned about the risks posed by international trade in hazardous wastes. This trade is addressed, in part, by the Basel Convention on the Control of Trans boundary Movement of Hazardous Wastes and Their Disposal. The convention came into force in 1992, but has yet to be ratified by the United States.

Trade liberalization makes it more likely that comparable goods produced under different environmental conditions will compete directly for market share. Environmentalists are concerned that trade liberalization may encourage countries to set low levels of environmental protection— not only standards, but their enforcement—to reduce production costs and encourage foreign investment. This, in turn, may force other countries to lower their environmental standards to maintain the competitiveness of their exports.

For most environmental problems, the evidence indicates that environmental expenditures have a minor impact on international competitiveness. However, the impact of, say, a large carbon tax to reduce greenhouse gas emissions could be very significant if imposed by some countries and not by others. Some suggest that countries with stringent environmental standards (primarily developed countries) should impose duties on imports from countries with not so stringent standards (primarily developing countries) to compensate for differences in expenditures on environmental protection.

Should countries of the North and South have comparable standards to protect the environment, health, and safety? Sustainable development means, among other things, remaining "within the carrying capacity of supporting ecosystems"; carrying capacity varies from country to country, depending on climate, geography, and other factors, and ecosystems may span national boundaries. Countries also differ in the resources they have available for environmental protection and the priority they assign it compared to, say, improved nutrition.

Should a distinction be made between standards that protect environmental carrying capacity, which varies from country to country, and standards that protect human health and safety?

The GATT recognizes the urgency of raising the living standards of developing countries and of the "progressive development" of their economies. It promotes increased market access under favorable conditions for their processed and manufactured products. It asks contracting parties to expand trade with developing countries by harmonizing and adjusting "national policies and regulations ... [and] technical and commercial standards affecting production."

The Uruguay Round of GATT negotiations have been underway since 1986. A "Draft Final Act Embodying the Results of the Uruguay Round" has been prepared, but has not yet been agreed upon. If concluded successfully, the Uruguay Round may improve market access for developing countries in such areas as natural resource-based products, tropical products, agriculture, and textiles and clothing. The North American Free Trade Agreement (NAFTA) between Mexico, the United States, and Canada was signed in December 1992 but has not yet been ratified by Congress. An environmental review of the draft agreement was performed by the United States (with EPA participation) and utilized in negotiations.

NAFTA provides that obligations of international environmental agreements on stratospheric ozone depletion, hazardous wastes, and endangered species take precedence over NAFTA obligations, subject to certain conditions.

Promotion of sustainable development is one of the NAFTA's stated purposes. The NAFTA confirms that each country may set standards to achieve the level of environmental protection it deems appropriate.

It discourages countries from relaxing environmental standards in order to attract investment. Important environmental issues not addressed in the NAFTA itself are being taken up in parallel and follow-up mechanisms, which include a pollution control program for the U.S.-Mexico border, a North American commission on the environment, and bilateral environmental agreements.

A number of options for reconciling trade and environmental interests are being studied and tested. The NAFTA and the parallel and follow-up mechanisms may offer a practical model for pursuing more open trade in concert with environmental protection.

UNCTAD is developing case studies on trade-environment linkages in Brazil, Colombia, India, the Philippines, and Turkey and is examining the environmental impact of producing and processing commodities such as cocoa, coffee, and rice. The GATT's Working Group on Environmental Measures and International Trade is investigating the trade impacts of environmental regulations for product packaging and labeling, an issue of particular concern to developing countries. An OECD working group is developing guidelines for improving the compatibility of trade and environmental policies.

An especially promising option is for countries to conduct environmental reviews of trade agreements and trade reviews of environmental agreements as a standard procedure. Such reviews, conducted early in the negotiating process, should help reveal the environmental and economic implications of these agreements and foster public comment and debate on the issues that are revealed.

As to environmental risks from traded products, it has been suggested that countries adopt international environment, health, and safety standards.

However, such standards, where they exist, may not provide an acceptable level of protection for countries that now have stringent domestic standards. One way to reduce trade barriers posed by product standards is to expand international cooperation on product risk assessment and testing, perhaps using as a model the OECD's cooperative program of investigation on chemicals and pesticides.

This approach might or might not lead to a larger and more acceptable set of international environmental standards, but it would make it easier to distinguish between legitimate standards and "green" protectionism. With respect to environmental risks from production processes, one alternative to trade restrictions is international cooperation on environmental labeling of traded products to reflect their life-cycle environmental impact, and thus influence consumer choice. Another option is to provide training and financial/technical assistance to exporting countries to help them change processes that cause environmental harm.

Promotion of trade in environmentally cleaner technologies deserves special attention, as does encouraging multinational corporations to apply their home country standards (if they are more stringent) to their operations in developing countries. As mentioned earlier, these options may work best if trade restrictions are available as a backup.

Finally, countries of the North and South could take concrete steps "to promote the internalization of environmental costs" in the prices of traded products, in keeping with Principle 16 of the Rio Declaration. Progress toward this goal would be a useful measure of the extent to which expanded trade contributes to sustainable development.

TRADE AND THE RIO DECLARATION

"States have, in accordance with the Charter of the United Nations and the principles of international law, the sovereign right to exploit their own resources pursuant to their own environmental and developmental policies, and the responsibility to ensure that activities within their jurisdiction or control do not cause damage to the environment of other States or of areas beyond the limits of national jurisdiction." (Principle 2)

"States should cooperate to promote a supportive and open international economic system that would lead to economic growth and sustainable development in $11 countries, to better address the problems of environmental degradation. Trade policy measures for environmental purposes should not constitute a means of arbitrary or unjustifiable discrimination or a disguised restriction on international trade." (excerpt from Principle 12)

"States should effectively cooperate to discourage or prevent the relocation and transfer to other State of any activities and substances that cause severe environmental degradation or are found to be harmful to human health." (Principle 14)

6 CHAPTER
Role of NGOs

The Stockholm Conference of UN on Human Environment in the year 1972 paved a way for various governments all over the world for setting up various departments and ministries on environment. It lead to various national and international commitments for the protection of environment and various protocols and agreements were signed after this conference.However, all these commitments and protocols have no sanctity as they were hardly being enforced. Lack of political will and scarcity of funds are the main reasons behind non or poor implementation of these agreements. Sometime it also happened that the decision taken in regard to environmental issues were not very sound hence their implementation was not visualized.

When such situations arouse, people who were environmentally conscious generated pressure either themselves or through some organizations which were not governed by the governments or the Non-Governmental Organizations(NGOs). The NGOs which are in the field of environmental management have rendered a commendable service. They exert all sorts of pressure on to national governments, the international agencies and the business corporations for fostering the cause of environment related issues.

The Non-Governmental Organizations of all over the world can be grouped in to following three categories:

- Agencies with restricted memberships which include scientific and professional members. These include International Council of Scientific Unions (ICSU); International Union for Conservation of Nature and Natural Resources (IUCN)
- Organizations involved mainly in collection of information, research and consultation. These include International Institute for Environment and The Institute of European Environment Policy etc.

- Organizations with open membership that are devoted to the cause of raising support and consensus for wildlife protection, improvement in the quality of environment, conservation of natural resources etc. These include Sierra Club, National Audubon Society, BNHS

The non-governmental Organisations have played a significant role at global level in developing lobbies especially for transboundary issues implying that environmental issues and problems are not limited within the national borders but are the problems of the entire humanity.

There are thousands of NGOs all over the world working in the various fields of environment. Some are working on pollution control, others on conservation of natural resources and some other are working in the development and poverty reducing programs.

Most of the NGOs usually work as mediators between governments and citizens. They work at different levels starting from grass root to national levels involving the various issuesand poor or socially disadvantaged people and provide them necessary support. There are NGOs which do not accept any aid from the government and raise funds through charities, however, some of them get government funding and also aid from other agencies.

Many NGOs like Greenpeace, Friends of earth and World Wide Fund (WWF) work at global level. They play a crucial role and sometimes put a lot of pressure even the governments had to change their policies. Greenpeace played a crucial role in preventing a company known as Shell to dump its worn out oil ship, the Brent Spar in North Sea.

In 1999 WTO organized a meeting in Seattle, Washington to develop agenda for next round of talks the Millenium Round. The meeting was attended by about 5000 delegates all over the world including the environmental ministers of various governments. The meeting was disrupted by thousands of protesters. The US authorities used tear gas to disburse them and arrested hundreds of protesters and imposed curfew. Finally the talks failed because the WTO was not willing to consider Environmental and poverty issues adequately.

Similarly, The Korean Federation of Environmental Movement opposed the move of Taiwan to buy a piece of land in North Korea to dump its nuclear waste.

In India, in February 1973, the Planning Commission approved a 240 MW Silent Valley Hydro Electric Project in Silent Valley, in Northern part of Kerala to harness the waters of Kunthipuzha. The dam when filled up would have submerged an area of 830 hectare out of which 500 hectare was a forested area. Harboring a unique ecosystem.

But a debate started in 1978 and it became a controversial issue. The Kerala Siksha Sahitya Parishad (KKSP) started Save Silent Valley campaign and KKSP fought to save forests not for their aesthetic or pure ecological value but for preserving a gene pool with its socio-economic implications. It related conservation issues to the need of the future generation and propagated a new ethic "We have not inherited the earth from our ancestors but have borrowed it from our children". Attempts were made by Friends of Trees and the Society for the Protection of Silent Valley.

The movement offered a unique foresight of people's participation, calling for right to information as an integral part of right to liberty.

The Chipko movement started in 1973 set another example of people's participation for the cause of environment and forests. People organised themselves in a group to save the forests. The people of the hills, specially the women literally embraced the trees when the loggers would come to fell them. Chipko movement is an example of how a non violent struggle by thousands of people can achieve the protection of environment. This movement also helped to seek for alternatives for meeting the needs of industry.

Similarly another example of non-governmental efforts for the cause of environmental protection and related issues is that of Narmada Valley Project for hydroelectric dam and irrigation project. The Narmada Bachao Andolan is fighting hard to save the ecology and people from being displaces from the area.

LIST OF SOME IMPORTANT NGOS

- International council of scientific unions (ICSU).
- InternationalUnion for Conservation of Nature and Natural Resources (IUCN)- Est. in 1948, H.Q.- Gland, Switzerland.
- World Wide Fund for Nature (WWF) (1961)- H.Q.- Gland, Switzerland.
- National Audubon Society (1880- in Auduon Park, NewYork).
- SIERRA CLUB (1892) San Francisco.
- BNHS-1883, Bombay.
- Centre for Science and Environment (CSE)-New Delhi.
- Scientific Committee on Problems of Environment (SCOPE)-1969.
- Centre for Environment and Education(CEE), Ahmadabad.

7

CHAPTER

Environmental Management

Environmental management is defined as a system that incorporates processes for summarizing, monitoring, reporting, developing and executing the environmental policies. The aim of encouraging an environmental management system is to ensure the healthy state of our planet for future generations. It also works towards preserving all forms of life.

In business, environmental management is defined as a corporate strategy that monitors, develops and implements environmental policies of an organization. It is a systematic approach that is gaining due prominence as consumers are looking for products and services that are eco-friendly and eco-aware.

The pattern of development which we have been following and the greed of the modern society to achieve every thing which nature has as a resource, have resulted in massive degeneration of our environment, depletion of natural resources and pollution of our environment, which is posing a threat to destroy the vital life support system of our planet earth. Entire humanity has not prospered by the exploitation of natural resources, two third of world's population still lives in shabby hutments, in wants, in misery and very poor unhygienic conditions.

To improve the lot of world's poor under-developed citizens, on one hand we have to accelerate economic development which will built up our capacity to protect the environment and, on the other hand this development should be conservation linked development or say it should be a sustainable development, i.e. a development which provides for the legitimate needs of present generation without comprising the ability of future generations to meet their own needs.

When we say environmental management we mean that to manage different types of problems which exist because of the deterioration of the conditions which are needed for the better living of human society and sustaining the natural processes.

Our objectives of environmental management should thus include following:

1. To provide for the legitimate needs of all people of the present generation and to bring them to a reasonable standards of life.
2. To manage economic development in such a way as to enhance and widen our resource base so that the ability of our future generation, which is likely to be more numerous than we are, to meet their own needs is not compromised.
3. To protect and preserve the environmental quality for the present and future generations.

ENVIRONMENTAL MANAGEMENT STRATEGIES

It has been the cumulative action, spread over several decades of so many people all over the world which has caused the degeneration of our natural resources and adverse changes in the environment of today.

- Though a large share of responsibility of bringing these changes goes to the industrialized countries of the world, but each locality, each region and each country has contributed to it in one way or other.
- Local or regional problems concerning environment as well as global issues which confront humanity to day have to be simultaneously addressed as it is many of the local problems only which add up to assume global dimensions.
- Emission of different gases from a small industrial unit, smoke coming out from a tiny power plant, automobile exhaust etc. while cause a local nuisance, but also add a little bit to the global burden of green house gases.
- Similarly, CFCs used and released from each small or large township while apparently causing no local problems could contribute a little to the ozone depleting substances of the stratosphere.
- Diverse means and methods are being used by environmentally conscious people all over the world to redress the environmental problems. But in general most of the efforts being carried out to protect the natural resources, wildlife, environment and building up a sustainable society and these include:
- Environmental Education.
- Sustainable use of natural resources.
- Checking environmental pollution and stabilizing global chemical cycles.
- Halting degeneration of biodiversity
- Placing a check on population growth.

OBJECTIVES OF ENVIRONMENTAL MANAGEMENT

The objectives of environmental management are as follows:

- Identifying environmental issues
- Finding solutions for environmental issues
- Establishing limits to avoid overuse
- Help to renew natural resources
- Minimize the use of natural resources
- Developing monitoring systems and research institutions
- Regeneration of degraded environment
- Review the environmental goals of an organization
- Setting environmental targets to minimize the environmental impact of an organization
- Control environmental pollution
- Ensuring environmental awareness program is being followed by every employee
- Review existing technologies and try to make then eco-friendly
- Make maximum utilization of natural resources
- Assess the impacts of potential activities on the environment
- Encourage resource conservation programs
- Develop strategies for improving the quality of life
- Implement ways for environmental protection
- Minimize the impact of natural disasters
- Identify, develop and implement policies related to sustainable development

1. Environmental Education

• Environmental education has a fundamental role to play in motivating people to adopt environmental friendly practices.

• We cannot expect people to act in appropriate way without an awareness of the problems. It causes, the impact on our daily life and the long range consequences.

• Many problems concerning environment and biosphere are simply there because so many people contribute little bits and pieces to it, all of which put together assume enormous dimensions.

• A little efforts, a little care exercised by each individual in the society could eliminate the entire problem. Starting from grass-root levels environmental education should involve all section of our society and should be able to:

(*i*) Create a general awareness concerning environmental problems.

(*ii*) Motivate people to conserve resources, protect environment and avoid extravagence.

(*iii*) Promote understanding and co-operation among people to face ecological issues.

(*iv*) Conserve indigenous knowledge, tradition and culture friendly to the environment.

(*i*) Creating a general awareness concerning environmental problems: A number of ecologically damaging activities could be stopped simply by providing adequate understanding of the consequences by involving to the people who indulge in such activities. Individuals, communities and human so as a whole should acquire the skill, will and experience to act singly as well as collectively to face the problem of environment.

(*ii*) Motivating people to conserve resources, protect environment and avoid extravagance: The need of the time is to preserve the normal functioning of the man dominated natural ecosystem in a healthy state. Like all other natural ecosystems, it should become self sustaining, *i.e.,* requiring none or nominal energy and material inputs while man should satisfy himself with whatever out put it generates without over straining itself. Whenever, there is an excess output it should be shared judiciously among the needy. It is the need of the time.

(*iii*) Promote understanding and co-operation among people to face ecological issues: Issue related to the environmental degradation are and shall continue to be ignored in favour of economic considerations as long as gross injustices in the pattern of distribution of natural resources prevail. The establishment of new economic order both at national and at international levels, which provides for the deprived and the needy, is necessary if we are to expect something concrete from the millions of inhabitants of tropical and subtropical countries of the world. There should be a free transfer of technology and means between the developed and developing countries of the world.

(*iv*) Conserving indigenous knowledge, tradition and culture friendly to the environment: Ancient traditions protect a number of plants and animal species and honour sacred grooves believed to be the homes of God.Much of this accumulated wisdom of past is to be saved

2. Sustainable Use of Natural Resources

Natural resources such as mineral wealth, fossil fuels, fertile soils, fresh water, live-stock and fisheries, wild life and forests etc. provide the mankind with everything which is needed directly or indirectly. These resources can be renewable or non-renewable. Renewable resources are those resources which can be regenerated whereas non-renewable resources are those resources which can not be regenerated once they are exhausted. Both the renewable resources and non- renewable resources have their own limitations.

Since a constant flow of materials is needed to maintain the loving beings, the society and economy, it is not one way consumption but a regular re-use and recycling which has so far enabled the biosphere to maintain it self since life began on this planet. Sustainable management of natural resources, therefore, involves different type of strategies for renewable and non renewable resources

***(a)* Sustainable management of non – renewable resources:** The sustainable management of our mineral wealth involves strategies to extend their life span which may be carried out by following :

1. Economy in the use of mineral resources, reduction in wastage and whenever possible the use of some cheaper substitute.
2. Re-use and recycling which shall reduce the demand on virgin material.
3. Making finished products that are long lasting which shall have the effect of decreasing the demand of the commodity, reducing thereby the demand of the metal concerned.
4. More efficient recovery of element from the deposits which shall curtail wastage and where smaller amounts of other elements are present in the deposits may also be tapped.
5. Protection of existing deposits and search for new ones.

***(b)* Sustainable management of renewable resources:** Sustainable management of our renewable resources involves:

- Avoiding over-exploitation and pollution of biotic systems: There is a limit up to which natural systems can provide resources to mankind. Withdraw beyond this limit tends to damage the productivity of the system.

  Both the affluence of developed countries as well as abstract poverty in under –developed countries is detrimental to the productivity of biotic systems. The root cause of environmental damage lies in want and misery of masses which inhabit developing countries of the world.
- Strengthening the resource base and augmenting regenerability of biotic systems:

The three important sectors which feed mankind are:

(*i*) Agricultural sector

(*ii*) Live-stock and Fisheries

(*iii*) Forest and wild life

***(i)* Agricultural Sector:** Agro-ecosystems which man sets up have to be provided fresh –water nutrients or fertilizers, chemicals to suppress insects and pests ad suitable hybrid varieties. The heavy reliance of modern agriculture on material energy inputs makes it unsuitable.

The best strategy for sustainable management of agro-ecosystems appears to be the promotion of a healthy microbial community which augments nutrient regeneration capacity, water holding potential and porosity of soils.

***(ii)* Live-stock and fisheries:** These provide to the mankind rich high quality proteins. For sustainable management of our livestock population we should adopt our traditional ways which served as means to convert useless organic matter left after the removal of grains from crop into a source of useful material and energy. The livestock raised on waste organic matter also serve the man to produce milk, meat, eggs, hide, woolens, manures and is a source of cheap labor. The livestock and poultry should be maintained as a useful nutrient loop between agricultural crop and man.

***(iii)* Forests and wild life**; Forests and wildlife provides a variety of useful products to the mankind. For thousands of people living in world's developing countries, resources of local forests and wild life products constitute an important source of livelihood.

In addition to these, forest are helpful in checking erosion and and degradation of soils, in shaping natural environmental and local climatic conditions by maintaining humidity, regulating temperatures impeding wind velocity and influencing precipitation. With destruction of forests natural habitats for wild life are lost. The use of Agricultural food resources with a little change in our food habits and a little adjustment these species may also be used to supplement the human food. This will widen our resources base and add sustainability to supplies by reducing the dependence on a few species.

3. Checking Pollution of Environment and Stabilizing chemicals Cycles

Ever since industrialization began, about two centuries ago mankind has altered earth's chemistry drastically which may have staggering ecological and economical consequences within our life times or those of our children.

Three important out-comes of the changes in earth's chemistry brought about by industrial societies are especially threatening and costly to the entire humanity.

These are :

1. Diminishing food security from a changing climate.
2. Degeneration of forests and green cover and reduction on biodiversity.
3. Risk to humans from exposure to chemical pollutants of the environment.

THE GRAVITY OF THE PROBLEM

A large number of questions remain unanswered for each of these threats and more research is urgently needed to clear the uncertainties surrounding them.

Waiting for a definite picture to emerge could, however, be dangerous as by the time we reach any conclusion, the world will already be committed and disastrous effects inevitable and irreversible. The wise strategy would be to avoid activities detrimental to the life support system right from the very beginning.

Since 1860, we have introduced about 200 billion tons of carbon into the atmosphere. Most of this carbon was earlier trapped in fossil fuels, limestone and organic matter.

The concentration of methane , another green house gas has roughly doubled science its 1600 A.D. level.

The amount of oxide of nitrogen added to the atmosphere by human activity roughly equals the total quantity of the NO_2 produced by all natural process combined. So is the quantity of sulphur dioxide added to the atmosphere which is about 100 million tons per year.

Lead, mercury, cadmium and Zinc etc. Are some to the metals which are being deposited at many places in concentrations toxic to man and other organisms.

MANAGEMENT STRATEGIES

In order to moderate the adverse impacts of human activity the following strategies are being followed:

1. Setting standards for effluents discharged in the environment.
2. Adopting more environment friendly technology.
3. Avoiding concentration of polluting industries and raising assimilative capacity of natural systems.

1. Setting standards for effluents discharged in the environment: This approach provides statuary standards for each polluting unit- be it an automobile or an industry. These standards specify permissible limits in terms of different parameters of effluent quality and the polluting unit is required to ensure that the effluent it discharge are within the specified limits.

These standards are based on evaluation of technical feasibility and economic viability. A regular monitoring of each relevant parameter is done in order to ensure that the discharges are within the prescribed limit. For example, waste water from Tannery should not have a BOD higher than 30 mg/l over 5 days period at 20 0 C. tanneries are required to treat their liquid effluent in such a way that BOD is reduced to levels below 30 mg/l over 5 day period at 20 °C.

Thus, management based on standards distributes the responsibility of reducing the load of pollution to the polluting units.

2. Adopting more environment friendly technologies: A much better way to restrict the pollution of environmental is adoption of more environment friendly technology. For example, of the three kinds of fossil fuels in common use these days, natural gas is the least polluting sources of energy.

3. Avoiding concentration of polluting industries and raising assimilative capacity of natural systems: we can enhance this capacity of the natural system by promoting green cover in areas surroundings the industry. Barren land should be treated properly, provided fertilizers and water so that it turns into an eutrophic system- capable of supporting suitable plants and a variety of microbes so as to serve as an effective instrument of assimilation.

PLACING A CHECK ON GROWTH OF HUMAN POPULATION

If we carefully mange our resources, ensure just and equitable distribution of necessities of life among all living being in this world, avoid extravagance, the planet has enough for all of us,. But the growth in population of one or two species – say for example, that of humans- cannot be indeterminate.

Howsoever, large may be the resources of a planet, whatsoever, may be its capacity for regeneration, indeterminate growth and consumption thereby, shall exhaust the carrying capacity of the system at some point of time in future and it shall collapse.

A line has to be drawn within which we shall have to confine our numbers.

DEVELOPMENT OF CONCEPT OF ENVIRONMENT MANAGEMENT SYSTEM

The concept of environmental management system (EMS) emerged in early nineties, though its genesis could be traced as back as in 1972 with the beginning of UN Conference on Human Environment and launching of United Nations Environment Program. (UNEP). Later the establishment of World Commission on Environment and Development(WCED), then came Montreal Protocol and Earth Summit in 1992.

All this instigated to evolve the concept of sustainable development and environment.

The International Standard Organisation developed ISO-14001 (Standard for environment management system) in 1996.

These standards address environmental concerns through the allocation of resources, assignment of responsibilities, ongoing evaluation of practices, procedures and processes and renders practical advice on implementation of such a system.

ISO 1400 SERIES

The ISO 1400 Environmental Management Standard series is a set of intentional standards that focus on environmental management. This series focuses, not on the actual product, but how the product is produced. Its origins can be traced all the way back to the United Nations Conference on Human Environment in 1972. It took another 20 years, however, before the standards and guidelines were actually put into place at the Rio Summit on Environment in 1992.

What is ISO 14001?

ISO 14001 is the most well-known standard in the ISO 1400 family. Unlike many other quality control standards, the ISO 14001 standard does not have any exact measures. Instead it serves as the framework of control for businesses and establishments to create their own Environmental Management Standards; it focuses on how the standards can be applied in a business or organization to meet the guidelines and standards. Each business must establish its own targets and performance measures.

The ISO 14001 Standard is also the only standard against which businesses and establishments can achieve certification from a third party. Achieving certification is based upon meeting all three of the components of the ISO 14001 Environmental Management Standards; minimizing how business operations negatively impact the environment, complying with the regulations and laws outlined in the EMS and continual improvement.

ISO 14001 Principles

The ISO 14001 Standard is based on the Plan-Do-Check-Act methodology, which is a system based on a concept of continual improvement. It encompasses a total of 17 elements that are grouped into five phases.

Environmental Policy (Plan): Review processes and products to identify the current elements of operation and how those elements impact the environment. Future operations are also assessed during the plan phase to determine how they may impact various environmental aspects. Impact may be direct (manufacturing process) or indirect (raw materials).

Planning (Do): Identify the resources that are required and document all procedures. Communication and participation are essential to ensure success, especially in top management positions.

Implementation and operation (Check): Measure and monitor processes. Report data and results.

Checking and corrective action (Act): Ensure that objectives are being met through a planned management review. Data gathered in step 3 is used to determine if any corrective action is needed. Make necessary changes.

Management review (Continual Improvement Process): Based on three dimensions that gradually move the business from operational environmental measurement towards a more strategic approach when dealing with environmental concerns and challenges. Dimensions include:

Expanding the Environmental Management Standards to more and more businesses areas

Enrichment by managing more and more processes, products, resources and activities

Upgrading to improve the organizational and structural framework of the EMS.

These are internationally harmonized standards relating to environmental management including eco-lebelling and environmental audit etc.

- ISO 14001: EMS, Specifications with guidance
- ISO 14004: EMS, General guidelines as principles, systems and supporting techniques
- ISO 14010-12: Guide lines for environmental auditing
- ISO 14014: Initial review
- ISO 14015:Environmental Site Assessment
- ISO 14020: Environmental Labeling
- ISO 14031: Evaluation of Environmental performance
- ISO 14040-14043: Life Cycle Assessment

- ISO 14050: Terms and guidelines
- Promote voluntary consensus approach
- Sensitize internal culture in organizations
- Minimize trade barriers
- Promote predictability and consistency of national standards
- Provide frame work to move beyond compliance, reduce liabilities and employees involvement
- Enhance public posture and customer trust
- Provide credibility and performance reporting
- Installs values, changes culture and improve public image
- provides an opportunity to link environmental objectives and targets with specific financial outcome thus ensuring that resources are made availble where they provide the most benefits in both financial and environmental terms.
- Such a system reduces liabilities of accidents
- Improves relations with regulatory bodies
- Improves government-industry relations

ENVIRONMENTAL IMPACT ASSESSMENT

Environmental impact assessment (EIA) is a tool used for decision making regarding developmental projects and programmes and it may be defined as a forma process use to predict the environmental consequences of any development project. EIA thus ensures that the potential problems are foreseen and addressed at an early stage in the project planning and design.

EIA is intended to identify the environmental, social and economic impacts of a proposed development prior to decision making. Hence using EIA it is possible to arrive at the following.

- The most environmentally suitable option to arrive at an early stage
- The Best practicable environmental option
- Alternative process

The project manager can them address these problems in order to avoid or minimize environmental impacts in conjunction with their project planning. This result in the likelihood of the project planning stages running smoother.

The environmental assessment is carried out by the developer although the task is often carried out by environmental consultants. Environmental assessment is carried out in order to produce an environmental statement / report. The environmental statement must include:

- A description of the project: location, design, scale, size etc.
- Description of significance effects
- Mitigating measures
- A non-technical summary

Different Types of Impact Assessment

Environmental impact assessment could encompass the following types of impact assessment.

- Climatic impact assessment
- Demographic impact assessment
- Development impact assessment
- Ecological impact assessment
- Economic and fiscal impact assessment
- Environmental auditing
- Environmental impact assessment
- Environmental management system
- Health impact assessment
- Project evaluation
- Public consultation
- Public participation
- Risk assessment
- Social impact assessment
- Strategic impact assessment
- Technological assessment

The Benefit of EIA

The following is a general overview of the many benefits offered by effective EIA:

- Reduced cost and time of project implementation
- Cost saving modifications in project design
- Increased project acceptance
- Avoiding impacts and violations of laws and regulation
- Improved project performance
- Avoiding waste treatment / cleanup

The benefits to local communities from taking part in environmental impact assessment include:

- A healthier local environment (forest, water sources, agricultural potential, recreational potential, aesthetic values, and clean living in urban areas)
- Improved human health
- Maintenance of biodiversity
- Decreased resource use
- Fewer conflicts over natural resource use
- Increased community skills, knowledge and pride

The EIA Process

The following two distinct stages in EIA:

1. **Preliminary assessment:** Carried out in the early stage of planning
2. **Detailed assessment:** carried out during project planning until the project plan is completed and is reported formally as an environmental impact statement.

The key elements of an EIA are the following:

1. **Scoping:** identify key issues and EIA is required based on information parties.
2. **Screening:** deciding whether an EIA is required based on information collected.
3. **Identifying and evaluating alternatives:** listing alternative sites and techniques and the impacts of each.
4. **Mitigating measures dealing with uncertainty:** review proposed action to prevent or minimize the potential adverse effects of the project.
5. **Issuing environmental statements:** reporting the findings of the EIA.

8 CHAPTER

Biofertilizers

One of the major concerns in today's world is the pollution and contamination of soil. The use of chemical fertilizers and pesticides has caused tremendous harm to the environment. An answer to this is the biofertilizer, an environmentally friendly fertilizer now used in most countries. Biofertilizers are organisms that enrich the nutrient quality of soil. The main sources of biofertilizers are bacteria, fungi, and cynobacteria (blue-green algae). The most striking relationship that these have with plants is symbiosis, in which the partners derive benefits from each other.

Plants have a number of relationships with fungi, bacteria, and algae, the most common of which are with mycorrhiza, rhizobium, and cyanophyceae. These are known to deliver a number of benefits including plant nutrition, disease resistance, and tolerance to adverse soil and climatic conditions. These techniques have proved to be successful biofertilizers that form a health relationship with the roots.

Biofertilizers will help solve such problems as increased salinity of the soil and chemical run-offs from the agricultural fields. Thus, biofertilizers are important if we are to ensure a healthy future for the generations to come.

MYCORRHIZA

Mycorrhizae are a group of fungi that include a number of types based on the different structures formed inside or outside the root. These are specific fungi that match with a number of favourable parameters of the the host plant on which it grows. This includes soil type, the presence of particular chemicals in the soil types, and other conditions.

These fungi grow on the roots of these plants. In fact, seedlings that have mycorrhizal fungi growing on their roots survive better after transplantation and grow faster. The fungal symbiont gets shelter and food from the plant which, in turn, acquires an array of benefits such as better uptake of phosphorus, salinity and drought tolerance, maintenance of water balance, and overall increase in plant growth and development.

While selecting fungi, the right fungi have to be matched with the plant. There are specific fungi for vegetables, fodder crops, flowers, trees, etc.

Mycorrhizal fungi can increase the yield of a plot of land by 30%-40%. It can absorb phosphorus from the soil and pass it on to the plant. Mycorrhizal plants show higher tolerance to high soil temperatures, various soil- and root-borne pathogens, and heavy metal toxicity.

LEGUME-RHIZOBIUM RELATIONSHIP

Leguminous plants require high quantities of nitrogen compared to other plants. Nitrogen is an inert gas and its uptake is possible only in fixed form, which is facilitated by the rhizobium bacteria present in the nodules of the root system. The bacterium lives in the soil to form root nodules (*i.e.* outgrowth on roots) in plants such as beans, gram, groundnut, and soybean.

BLUE-GREEN ALGAE

Blue-green algae are considered the simplest, living autotrophic plants, i.e. organisms capable of building up food materials from inorganic matter. They are microscopic. Blue-green algae are widely distributed in the aquatic environment. Some of them are responsible for water blooms in stagnant water. They adapt to extreme weather conditions and are found in snow and in hot springs, where the water is 85 °C.

Certain blue-green algae live intimately with other organisms in a symbiotic relationship. Some are associated with the fungi in form of lichens. The ability of blue-green algae tophotosynthesize food and fix atmospheric nitrogen accounts for their symbiotic associations and also for their presence in paddy fields.

Blue-green algae are of immense economic value as they add organic matter to the soil and increase soil fertility. Barren alkaline lands in India have been reclaimed and made productive by inducing the proper growth of certain blue-green algae.

BIOREMEDIATION

Enormous quantities of organic and inorganic compounds are released into the environment each year as a result of human activities. In some cases these releases are deliberate and well regulated (*e.g.* industrial emissions) while in other cases they are accidental (*e.g.* chemical or oil spills). Petroleum and its products are one of the most common environmental pollutants. They are a fire hazard, threat to marine life, and a source of air and groundwater pollution. They contaminate land and water bodies by accidental spills like the Alaska Oil spill in 1989 and oil spills during the Gulf War, leakage from pipelines, and other human activities. Detoxification of the contaminated sites is expensive and time consuming by conventional chemical or physical methods.

Bioremediation consists of using naturally occurring or laboratory cultivated micro-organisms to reduce or eliminate toxic pollutants. Petroleum products are a rich source of energy and some organisms are able to take advantage of this and use hydrocarbons as a source of food and energy. This results in the breakdown of these complex compounds into simpler forms such as carbon dioxide and water. Bioremediation thus involves detoxifying hazardous substances instead of merely transferring them from one medium to another. This process is less disruptive and can be carried out at the site which reduces the need of transporting these toxic materials to separate treatment sites.

Using bioremediation techniques, TERI has developed a mixture of bacteria called 'oilzapper' which degrades the pollutants of oil-contaminated sites, leaving behind no harmful residues. This technique is not only environment friendly, but also highly cost-effective.

BT COTTON

Cotton and other monocultured crops require an intensive use of pesticides as various types of pests attack these crops causing extensive damage. Over the past 40 years, many pests have developed resistance to pesticides.

So far, the only successful approach to engineering crops for insect tolerance has been the addition of Bt toxin, a family of toxins originally derived from soil bacteria. The Bt toxin contained by the Bt crops is no different from other chemical pesticides, but causes much less damage to the environment. These toxins are effective against a variety of economically important crop pests but pose no hazard to non-target organisms like mammals and fish. Three Bt crops are now commercially available: corn, cotton, and potato.

As of now, cotton is the most popular of the Bt crops: it was planted on about 1.8 million acres (728437 ha) in 1996 and 1997. The Bt gene was isolated and transferred from a bacterium bacillus thurigiensis to American cotton. The American cotton was subsequently crossed with Indian cotton to introduce the gene into native varieties.

The Bt cotton variety contains a foreign gene obtained from bacillus thuringiensis. This bacterial gene, introduced genetically into the cotton seeds, protects the plants from bollworm (A. lepidoptora), a major pest of cotton. The worm feeding on the leaves of a BT cotton plant becomes lethargic and sleepy, thereby causing less damage to the plant.

Field trials have shown that farmers who grew the Bt variety obtained 25%–75% more cotton than those who grew the normal variety. Also, Bt cotton requires only two sprays of chemical pesticide against eight sprays for normal variety. According to the director general of the Indian Council of Agricultural Research, India uses about half of its pesticides on cotton to fight the bollworm menace.

Use of Bt cotton has led to a 3%–27 increase in cotton yield in countries where it is grown.

DID YOU KNOW?

By the next generation the world population is estimated to double and the demand on food is also expected to grow relatively.

With the progress in biotechnology there will be less usage of chemicals. The new methods applied are making crops tolerant to insects, and to certain herbicides that will kill weeds and other pests.

In the early 1970s, when rice in South-East Asia was attacked by the grassy stunt virus, a gene from a wild strain of rice (Oryza nivara) from Uttar Pradesh helped save the crop.

Brassica (rapeseed-mustard) is the second most important edible oilseed crop in India after groundnut and accounts for nearly 30% of the total oilseeds produced in the country.

Many people who are vegetarians feel that if they consume vegetables that contain animal genes through the fertilizer that is applied, their food would not be purely vegetarian. This is a matter where each person has his/her own choice and opinion and this can differ from person to person.

Swiss researchers recently announced a major breakthrough in genetically modified rice grains to contain more Vitamin A and iron.

India has emerged as the world's largest producer of milk, the world's second largest producer of vegetables and fruits, and of wheat. It has even overtaken USA both in the production and per hectare yield of wheat.

DNA

Since the time Gregor Mendel began studying about inheritance in garden plants some 150 years back, researchers have worked to learn more about the language of life – how characteristics pass from one generation to another. Researchers began to understand DNA from the 1800s when they stated that all living beings, whether plants, humans, animals, or bacteria, comprised cells that have the same basic components.

Living organism are made up of cells, *i.e.* cells are the basic units of life. For example, each of us is made up of billions of this basic unit. If one closely inspects the structure of the cell, one is likely to find various smaller bodies or organelles like mitochondria that generates the energy required to perform all life processes ('the powerhouse'), chloroplast (only in green plants and responsible for their coloration), the central core – 'the nucleus, to name a few. The nucleus harbours the blueprint of life and the genetic material – DNA or deoxyribonucleic acid – and is the control centre of any cell. The genetic material or the blueprint is contained in all the cells that make up an organism and is transmitted from one generation to another. A child inherits half of the genetic material from each of his/her parents.

The chemical structure of everyone's DNA is the same. Structurally, DNA is a double helix: two strands of genetic material spiraled around each other. Each strand contains a sequence of bases, also called nucleotides. A base is one of four chemicals: adenine, guanine, cytosine, and thymine. The two strands of DNA are connected at each base. Each base will only bond with one other base, as follows: Adenine (A) will only bond with thymine (T), and guanine (G) will only bond with cytosine (C). If one strand of DNA looks like A-A-C-T-G-A-T-A-G-G-T-C-T-A-,the DNA strand bound to it will look like T-T-G-A-C-T-A-T-C-C-A-G-A-T-C.

Together, the section of DNA would be represented as given in Figure

T-T-G-A-C-T-A-T-C-C-A-G-A-T-C

A-A-C-T-G-A-T-A-G-G-T-C-T-A-G

The length of the DNA strand varies from organism to organism but within individuals of a particular species it is nearly constant. For example, a certain virus may have only 50 000 (5×10^4) bases constituting the genetic material whereas a human cell contains nearly 3.2 billion (3.2×10^9) bases in each of the cells (except the germ line cells). The amount and sequence in all the cells of an organism is identical. The DNA is for most part of the time present as condensed body called chromosomes (coloured body) except when it is replicating or dividing. A piece of a chromosome that dictates a particular trait, for example, eye and skin colour in humans, is called a gene. In any cell, the DNA can be classified into two categories – the sequence that codes for traits or genes and the sequence that has no apparent function or the non-coding DNA. The coding sequence (genes) in humans constitutes only five per cent of the total DNA and is identical in all humans. The non-coding sequence, which is nearly 95% in humans, varies from one individual to another, and forms the basis of DNA fingerprinting.

DNA FINGERPRINTING

The only difference between two individuals is the order of the base pairs. Each individual has a different sequence of DNA, specially in the non-coding region. Using these sequences, every person could be identified solely by the sequence of their base pairs. However, because the entire DNA is so huge, the task would be time-consuming and nearly impossible. Instead, scientists are able to use a shorter method.

The steps involved in DNA fingerprinting can be summarized as follows:

- Isolating the DNA in question from the rest of the cellular material in the nucleus.
- Cutting the DNA into several pieces of different sizes.
- Sorting the DNA pieces by size. The process by which the size separation, or 'size fractionation', is done is called gel electrophoresis.
- This is the basic concept behind fingerprinting technique.

DNA FINGERPRINTING IN PLANTS

The concept of DNA fingerprinting can also be extended to plants and many institutions in the country are doing it today. TERI has successfully generated fingerprints of various medicinal plants such as neem, ashwagandha, and amla with the objective of determining their identity. With the help of fingerprints one can find out the genetic diversity in India. This knowledge has profound implications. Based on the extent of genetic diversity, one can establish the centre of origin of a particular plant species. And having done that we are better equipped to prevent bio-piracy or the theft of our genetic resources.

THE GREEN REVOLUTION

The world's worst recorded food disaster occurred in 1943 in British-ruled India. Known as the Bengal Famine, an estimated 4 million people died of hunger that year in eastern India (which included today's Bangladesh). Initially, this catastrophe was attributed to an acute shortfall in food production in the area. However, Indian economist Amartya Sen (recipient of the Nobel Prize for Economics, 1998) has established that while food shortage was a contributor to the problem, a more potent factor was the result of hysteria related to World War II, which made food supply a low priority for the British rulers.

When the British left India in 1947, India continued to be haunted by memories of the Bengal Famine. It was therefore natural that food security was one of the main items on free India's agenda. This awareness led, on one hand, to the Green Revolution in India and, on the other, legislative measures to ensure that businessmen would never again be able to hoard food for reasons of profit.

The Green Revolution, spreading over the period from1967/68 to 1977/78, changed India's status from a food-deficient country to one of the world's leading agricultural nations. Until 1967 the government largely concentrated on expanding the farming areas. But the population was growing at a much faster rate than food production. This called for an immediate and drastic action to increase yield. The action came in the form of the Green Revolution. The term 'Green Revolution' is a general one that is applied to successful agricultural experiments in many developing countries. India is one of the countries where it was most successful.

There were three basic elements in the method of the Green Revolution:

- Continuing expansion of farming areas
- Double-cropping in the existing farmland
- Using seeds with improved genetics.

The area of land under cultivation was being increased from 1947 itself. But this was not enough to meet the rising demand. Though other methods were required, the expansion of cultivable land also had to continue. So, the Green Revolution continued with this quantitative expansion of farmlands.

Double cropping was a primary feature of the Green Revolution. Instead of one crop season per year, the decision was made to have two crop seasons per year. The one-season-per-year practice was based on the fact that there is only one rainy season annually. Water for the second phase now came from huge irrigation projects. Dams were built and other simple irrigation techniques were also adopted.

Using seeds with superior genetics was the scientific aspect of the Green Revolution. The Indian Council for Agricultural Research (which was established by the British in 1929) was reorganized in 1965 and then again in 1973. It developed new strains of high yield variety seeds, mainly wheat and rice and also millet and corn.

The Green Revolution was a technology package comprising material components of improved high yielding varieties of two staple cereals (rice and wheat), irrigation or controlled water supply and improved moisture utilization, fertilizers, and pesticides, and associated management skills.

BENEFITS

Thanks to the new seeds, tens of millions of extra tonnes of grain a year are being harvested.

The Green Revolution resulted in a record grain output of 131 million tonnes in 1978/79. This established India as one of the world's biggest agricultural producers. Yield per unit of farmland improved by more than 30% between1947 (when India gained political independence) and 1979. The crop area under high yielding varieties of wheat and rice grew considerably during the Green Revolution.

The Green Revolution also created plenty of jobs not only for agricultural workers but also industrial workers by the creation of related facilities such as factories and hydroelectric power stations.

SHORTCOMINGS

In spite of this, India's agricultural output sometimes falls short of demand even today. India has failed to extend the concept of high yield value seeds to all crops or all regions. In terms of crops, it remains largely confined to foodgrains only, not to all kinds of agricultural produce.

In regional terms, only the states of Punjab and Haryana showed the best results of the Green Revolution. The eastern plains of the River Ganges in West Bengal also showed reasonably good results. But results were less impressive in other parts of India.

The Green Revolution has created some problems mainly to adverse impacts on the environment. The increasing use of agrochemical-based pest and weed control in some crops has affected the surrounding environment as well as human health. Increase in the area under irrigation has led to rise in the salinity of the land. Although high yielding varieties had their plus points, it has led to significant genetic erosion.

Natural Resources and Associated Problems

CONSERVATION

The expansion of population at a generatric rates and one side expansion of technology to achieve certain goals has lead to the control between human beings and environment. The population technology and land use contribute to the environment crisis. Conservation may be defined as the most efficient and most beneficial utilization of natural resources.

In words of **Raymond F. Dasimon** of IUCN conservation is the rational us of environment to provide a high quality of living for the mankind. According to him, the most important aspect of conservation would be the maintenance or development on earth of the widest possible diversity.

On the basis of their abundance and availability natural resources can be classified in to two categories:

1. Renewable Resources: Resources that have the inherent capacity to reappear, or replenish themselves by quick recycling, reproduction, and replacement within a reasonable time and maintain themselves, are called renewable resources. *e.g.*, Soil, water, living organism or wildlife, aquatic organism, Forests.

OR

It is a natural resources which gets replenished, recycled or reproduced and should not be used beyond its renewability.

2. Non-renewable Resources: Resources that lack the ability for recycling and replacement are called non-renewable resources. *e.g.*, fossil fuels like coal, petroleum, and minerals.

OR

It is a natural resources which is likely to get exhausted with use because of its lack of recycling (*e.g.*, fossil fuels) or very long recycling time (*e.g.*, minerals).

Renewable resources can become non-renewable if used too rapidly by improper management.

FOREST RESOURCES

A forest is the most effective ecosystem that supports many kinds organism and absorbs the energy of the sun to build up organic materials. The world watch institute in its "state of the world 1988" report revealed that Indian forest cover declined from 16.9% in early 1970's to 14.1% in the early 1980's, an average loss of 1.3 million hectare per year. The forest cover within 100 kilometer of India's Major cities has dropped by 15% in the less than a year.

According to Brewabaker 1984 the total forest area of the world in 1900 was nearly 7000 million hectare by 1975 it was reduced to 2890 M hec. Its estimated that by 2000 A.D. It would be merely 2370 M hec. The major reduction will be in torpics and sub-tropics (40.2%) being only 0.6% in temperate areas.

According to Central forest commission (C.F.C.) 1980, the forest cover was around 74.8 M hec. Among 16 different forest types of the country the most common is tropical dry deciduous (38.1%), followed by tropical moist deciduous forest (30.9%).

The other forest types are as follows:

- Tropical thorn – 6.9%
- Tropical dry ever green – 0.1%
- Pure coniferous (High mountainous areas) – 6.3%
- Sal forest – 16%
- Teak forest – 13%
- Broad leaved (excluding sal and teak) -55.8%
- Bamboo's (Including plantation) – 8.8%

Out of the total Sal forest area, nearly 96% is owned by the government, 2.6% by the corporate bodies and rest are in private ownership. The exponential growth in the population has resulted a decrease in the forest cover though traditionally Indians have a deep involvement with forest and forestry.

The national forest policy 1952 enunciated that 1/3 of the geographic areas of the country should be under to forest cover. However, there had been continuous deforestation in the country for various regions and it's estimated that 4.238 M hec. Forest land was officially diverted for non-forest purposes between 1951-52 and 1979-80.

The government of India enacted the forest (conservation) act, 1980 with view to check indiscriminate – devastation and diversion of the forest land to non forest land or non forest purposes.

The forest (conservation) act 1980 was amended in 1988 to incorporate strict panel provisions against the violators of the act. The growing demand of cities has destroyed our forests. Besides these local cattle, goats, sheep etc. not only destroy the vegetation but also fell, cut the roots of the vegetation.

India today is poorest in the world as per capita land is concerned. The per capita forest land in India 0.10 hectare compared to the world average of 1.00 hectare, Canada 11.2 hec, Australia 7.6 hec. India is losing 1.50 M hec, of forest cover each year and if trend continuous in next 20 years or so we may reach to zero forest in our country.

USE OF FORESTS

Commercial Uses

Forest provides us a large number of commercial goods which include timber, fire wood, pulp wood, food items, gum, resins, nonedible oils, minor forest products etc. One third of the wood harvested is used for paper. Many forested are also used for mining, grazing, agriculture and recreation and for development activities.

Over Exploitation of Forests

The history of the exploitation of forest is as gold as man himself, but during earliere times it was balanced through a natural growth process because at that time forest cutting was done for personal or community use only. But with the time expansion of agriculture, forestlands have been cleared.

During the colonial period commercial exploitation use of forests nowadays has reached such and extent that it has become a threat to the environment in the form of increase in temperature, lesser precipitation, increased rate of soil erosion, increase in frequent and volume of floods, loss of soil productivity, imbalance of ecosystem etc.

The harmful effects of deforestation are so much that all over the world people and authorities have realized that forest resources must be conserved property in order to protect the ecosystem. The forest is a national resources and a social asset. It yields a great social profit which lies wholly outside the realm of business. But, at present, most of the forests of the world are so used that experts predicts dire calamities in not-too-distant future and irreparable damage on a catastrophic scale. If properly used and put on a sustained yield basis, it will be one of man's greatest resource and for this conservation of forest is the only alternatives.

Forest Conservation

The following steps should be taken for the conservation of forests:

(*a*) Regulated and planned cutting of trees: One of the main reasons of deforestation is commercial felling of trees. According to an estimate, about 1,600 million cubic meters of wood have been used for various purposes in the world. Although trees are considered as perennial resource, when exploited on a very large scale, their revival cannot be possible. Therefore, cutting should be regulated by adopting methods like clear cutting, selective cutting and shelter wood cutting.

The clear cutting method is useful for those areas where the same types of trees are available over a large area. In that case, trees of same age groups can be cut down in a selective are and then marked for re-plantation. **In selective cutting** only mature trees are selected for cutting. This process is to be followed in rotation. **Shelter wood cutting** is where first of all useless trees having been cut down followed by medium and best quality timber trees. The time gap between these cuttings is helpful in re-growth of trees. In regulated cutting only one – third of the forest area is selected for use and rotational system is always followed by for their protection. The forest can be managed in such a way that a timber crop may be harvested indefinitely year after year without being depleted. This technique is called the 'sustained yield' method adopted by many countries of the world.

(*b*) Control over forest fire: Destruction or loss of forest by fire is fairly common, because trees are highly exposed to fire and once started it becomes difficult to control, sometimes, the fire starts by natural process, i.e. by lightning or by friction between trees specially during speedy winds, while in most of cases it is also by man either intentionally or un-intentionally. According to an estimate, during the period from 1940 to 1950, in the US alone, fires consumed an average of 21.5 million acres of timber yearly and as many as 1,175,664 cases of forest fires occurred during 1955 to 1964 period.

In order to save forest form fire it is necessary to adopt latest techniques of firefighting. Some of the fire suppression techniques are to develop three meter wide fire lanes around the periphery of the fire, back fires, arrangement of water spray, fire retardant chemicals should be sprayed from back tank and if possible by helicopters. There must be a trained staff of fire fighters to control the fire.

(*c*) Reforestation and afforestation: The sustained yield concept reveals that whenever timber is re-moved, either by block cutting or by selective cutting, the denuded area must be reforested. This may be done by natural or artificial methods. Similarly, any forested land which has been destroyed by fire or mining activities should be reforested. In rugged terrain aerial seeding is the method of choice. Besides all this, fresh afforestation program should be started. New plantations will not only increase the forest cover but also help in making up the eco-balance. For afforestation, selection of trees should be done according to local geographical conditions and care must be taken during initial growth of the trees.

(*d*) Protection of forest: The existing forests should be protected. Aprart from commercial cutting, unorganized grazing is also one of the reasons. There are several forest diseases resulting from parasitic fungi, rusts, mistletoes, viruses and nematodes which cause the destruction of trees. The forests should be protected either by use of chemicals spray, antibiotics or by development of disease resistant strains of trees.

(*e*) Role of government in forest conservation: Both national and provincial governments can take some steps in the direction of conservation of forests, such as:

1. Pass acts for the conservation of forests
2. Survey of the forest resources
3. Categorized of forest areas and proper delimitation of reserved forest areas.
4. Find out the areas where reforestation can be done
5. Regulate the commercial use of forest products
6. Protect forest form fire, mining and other natural calamities
7. Develop national parks
8. Encourage forests developments activities like social forestry, agro-forestry etc.
9. Prepare master plan, both for long term and short term period, etc.

WATER RESOURCES:

Water resources are sources of water that are useful or potentially useful to humans. It is important because it is needed for life to exist. Many uses of water include agricultural, industrial, household, recreational and environmental activities. Virtually all of these human uses require fresh water. Only 2.5% of water on the Earth is fresh water, and over two thirds of this is frozen in glaciers and polar ice caps. Water demand already exceeds supply in many parts of the world, and many more areas are expected to experience this imbalance in the near future.

Water which is an essential element of life is found in a number of springs and perennial rivers which have their origin in the Himalayas in the north-east India and other mountain ranges in other country parts of the country. The springs, though extensively distributed and provide drinking water for a fairly large no. of villages, situated on high hill tops, are not, so far as is known, large enouth to permit their scientific management and utilization for any large scale drinking water supply scheme. However, a limited location such as in Kashmir, large springs proves copious discharges and can be developed for drinking water supply. In any scheme of preserving and developing the Himalayan environment, the detailed studies of springs can yield fruitful results.

A large number of perennial river systems originate and flow down the Himalayan slopes and these provides a very large potential for the development of drinking water and irrigation supply and for the generation of Hydro-electric power. While implementing these developmental schemes, however, it is necessary to ensure that adverse environmental impacts do not arise. The Himalayas play an important role in replenishing the water resources of our country. This is by their role and acting as a barrier range for the monsoon winds and thus enabling the precipitation of their moisture as snow or rain on the southern slopes. The resulting run of not only feeds the Himalayan drainage system but also recharges many of the aquifers, laying in the Tarai Bhabar and Indo-Gangetic plains, Thus spreading the benefits over a very large are and to a large segment of our population. In providing large storage in the Himalayan river system for permitting irrigation and power development, the efforts should be made to see that the ultimate benefits occurring by way of ground water recharge and supplementation of ground water storage are not adversely affected, there by leading to a degradation of the human environment in these places.

The Himalayan glacier also represents a large reservoir of fresh water resources. The total number of glaciers in the Himalayas is about 15,000 Km. Von Wissmen has estimated that the glaciers represent an ice cover of the order of 33,000 Km^2 representing nearly 1,400 representing nearly 1, 400 Km^3 of ice. So far only estimates of this resource were available but in recent years, the geographical division of G.S.I. located at Lucknow and Calcutta has taken up systematic and scientific studies of the Himalayan glaciers. These studies are valuable for the management of water resources and for monitoring the environmental changes brought about by the glacier in term of temperature and pattern of wind distribution.

Oceans: The oceans on the earth cover some 362 million Km^2 area, and are the sink into which all water finally flows. The water flowing over the land picks up a number of elements and compounds from the soil and air, and transports these substances in solution and deposits them in the ocean. Since there is no exit for these dissolved substances, they tend to accumulate in sea water. The average total salt content of the ocean is 35% (35 g per kg of water). The most concentrated substances in the ocean are sodium chlorides, which makes nearly 90 % of the total dissolved salts. Sea water contains about 57 elements in solution.

Ocean cannot be considered primary, resources for the direct water supply for domestic, agriculture or industrial are owing to their high salt content. However, at some places, the ocean water is being used for these purposes after the process of desalination. Oceans are very useful resources for obtaining other commodities like foods (fish and other sea food), common salt and other such items. In certain areas, the electrical energy is also obtained from the tidal energy of the oceans.

Ground water: The volume of ground water is much greater than that of all fresh water lakes and streams combined. Underground water plays an important role in the overall water balance of the environment. As a reservoir, it has an enormous capacity to store water in rainy periods, which can be utilized during dry periods. It helps controlling the river flow. Ground water is a primary source of freshwater in several towns and rural areas. The existing utilization of ground water in India is currently estimated to 45,000 million m^3. It is widely used as a source of water for irrigation and for other farm uses. The contribution of ground water to total irrigation is about 40% in India. Because ground water supplies are recharged by atmospheric precipitation, problems may result if the amount withdraw from ground water is greater that replenished by rainfall.

USES OF FRESH WATER

Uses of fresh water can be categorized as consumptive and non-consumptive (sometimes called "renewable"). A use of water is consumptive if that water is not immediately available for another use. Losses to sub-surface seepage and evaporation are considered consumptive, as is water incorporated into a product (such as farm produce). Water that can be treated and returned as surface water, such as sewage, is generally considered non-consumptive if that water can be put to additional use.

AGRICULTURAL

It is estimated that 70% of world-wide water use is for irrigation. In some areas of the world irrigation is necessary to grow any crop at all, in other areas it permits more profitable crops to be grown or enhances crop yield. Various irrigation methods involve different trade-offs between crop yield, water consumption and capital cost of equipment and structures. Irrigation methods such as most furrow and overhead sprinkler irrigation are usually less expensive but also less efficient, because much of the water evaporates or runs off. More efficient irrigation methods include drip or trickle irrigation, surge irrigation, and some types of sprinkler systems where the sprinklers are operated near ground level. These types of systems, while more expensive, can minimize runoff and evaporation. Any system that is improperly managed can be wasteful. Another trade-off that is often insufficiently considered is salinization of sub-surface water.

Aquaculture is a small but growing agricultural use of water. Freshwater commercial fisheries may also be considered as agricultural uses of water, but have generally been assigned a lower priority than irrigation (see Aral Sea and Pyramid Lake).

As global populations grow, and as demand for food increases in a world with a fixed water supply, there are efforts underway to learn how to produce more food with less water, through improvements in irrigation methods and technologies, agricultural water management, crop types, and water monitoring.

INDUSTRIAL

A Power Plant in Poland

It is estimated that 15% of world-wide water use is industrial. Major industrial users include power plants, which use water for cooling or as a power source (i.e. hydroelectric plants), ore and oil refineries, which use water in chemical processes, and manufacturing plants, which use water as a solvent.

The portion of industrial water usage that is consumptive varies widely, but as a whole is lower than agricultural use.

HOUSEHOLD

Drinking Water

It is estimated that 15% of world-wide water use is for household purposes. These include drinking water, bathing, cooking, sanitation, and gardening. Basic household water requirements have been estimated by Peter Gleick at around 50 liters per person per day, excluding water for gardens.

Most household water is treated and returned to surface water systems, with the exception of water used for landscapes. Household water use is therefore less consumptive than agricultural or industrial uses.

Recreation

Recreational water use is a very small but growing percentage of total water use. Recreational water use is mostly tied to reservoirs. If a reservoir is kept fuller than it would otherwise be for recreation, then the water retained could be categorized as recreational usage. Release of water from a few reservoirs is also timed to enhance whitewater boating, which also could be considered a recreational usage. Other examples are anglers, water skiers, nature enthusiasts and swimmers.

Recreational usage is non-consumptive. However it may reduce the availability of water for other users at specific times and places. For example, water retained in a reservoir to allow boating in the late summer is not available to farmers during the spring planting season. Water released for whitewater rafting may not be available for hydroelectric generation during the time of peak electrical demand.

Environmental

Explicit environmental water use is also a very small but growing percentage of total water use. Environmental water usage includes artificial wetlands, artificial lakes intended to create wildlife habitat, fish ladders around dams, and water releases from reservoirs timed to help fish spawn.

Like recreational usage, environmental usage is non-consumptive but may reduce the availability of water for other users at specific times and places. For example, water release from a reservoir to help fish spawn may not be available to farms upstream.

Mineral Resources

The world mineral has been derived from *minera,* meaning an ore or specimen, which shows that it dates back to the beginning of mining. In fact mineralogy is a branch of geological science which deals with the study of earth's crust. A mineral may be defined as a homogenous substance which has a specific chemical composition and is produced by natural inorganic process. Although widely varying in nature, and abundance, from common elements such as iron to rare earth such as titanium, they are critical to most of the goods and transportation and energy related products used in modern societies. Most of the mineral have definite chemical and physical properties which are related to the chemical composition and crystal structure of their constituents. The artificial materials, which are, infact not entirely of natural origin are not considered as minerals.

Any naturally occurring concentration of a free element or compound of two or more elements in solid form is called a mineral deposit. Although a few minerals, such as gold and silver, occur as free elements, most are found as various compounds of only eight elements that make up 99.3% earth's crust. A mineral deposit with a high enough concentration of at least on metallic element to permit it to be extracted and sold at a profit is called an ore or ore deposit.

An ore that contents a relatively large concentration of a desired metallic element is called a high grade ore; one with a relatively low concentration is known as a low grade ore. Estimating how many a particular non renewable mineral resources exists one earth and how much of it may be located an extracted at an affordable price is a complex controversial process. The term total resource refers to the amount of a particular material that exists on earth. The future availability of mineral resources depends not only on its actual or potential supply but also on how rapidly this supply is being depleted to the point where extracting and processing what remains is to costly. Five country of the world the USA, Canada, Australia, South Africa, Russia supply the world with most of 20 minerals that make up 98% of all known fuel minerals consumed in the world.

Depletion time is the length of time it makes to use up certain proportion usually 80% of the reserves of a mineral at a given great of use. A traditional measures of the projected availability of no renewable resources is the reserve to production ratio – the number of years proven reserves of particular nonrenewable mineral will last at current annual production rate. The shortest depletion time assumes no recycling or reuse and no increases in reserves. A longer depletion time assume that recycling will stretch exiting reserves and the better meaning technology, higher prices, and new discoveries increases reserves. While world population doubled between 1950 and 1999, global production six key metals (Aluminum, Copper, Lead, Nickel, Tin, Zinc) increased more than eight folds during the same period, world reserves of copper increased, lead almost three fold, zinc four fold and aluminum almost nine fold.

Indian subcontinent is rich in both, renewable and nonrenewable resources. As in matter of fact, that mineral exploitation in India has started late and there are encouraging indication of established mineral potential of India is so far been very limited. The reason for this are not yet clearly understood. One of the reason may be the late start of exploration of resources in the Himalayan region. There is also a possibility that their may be a conceptual barrier in the search for mineral resources. The pre-Cambrian of the Himalayan mobile belt shows the subdued base metal mineralization as in the case crystalline of Kumaon, the salkhala rocks of Kashmir and thimpu crystalline of Bhutan. Likewise with the characteristic volcano sedimentary formation, the phenerzoic of Himalayas and the lime stone deposits of Kumaon Himalayas. The dolomite and lime stone deposits of Bhutan and Arunachal Pradesh, the gypsum deposits of Eastern Bhutan and the coal resources of Drjeeling, Sikkim, Bhutan and Arunachal Pradesh. The assumption that these resources can not be saved are now providing wrong and it is seriously being realized that a severe shortage will developed if necessary attention is not paid for their conservation.

Important Mineral Resources of India

Following are the important mineral reserves of India:

S.No.	Minerals	Reserves (Tonnes)	States
1.	Bauxite	252.53	Andhra pradesh, Bihar, Gujrat, Karnataka, Goa, Jammu and Kashmir, Maharashtra, Orissa, Tamil Nadu, U.P.
2.	Gold	88	Karnataka, Andhra pradesh
3.	Iron ore	1274	Andhra pradesh, Kerala, Tamil Nadu, Karnataka, Goa, Maharashtra, M.P., Orissa, Bihar
4.	Lead	16.75	Andhra pradesh, Tamil Nadu, U.P., Orissa, West Bengal, Sikkim, Gujrat, Rajasthan.
5.	Manganese	29.4	Maharashtra, Karnataka, Bihar, Orissa, Andhra Pradesh
6.	Nickel	29.4	Maharashtra, Karnataka, Orissa, Manipur, Nagaland
7.	Tungsten	2329	U.P., West Bengal, Rajasthan, Maharashtra, Karnataka
8.	Diamond	12.75	Andhra Pradesh, M.P.
9.	Dolomite	496.7	Orissa, M.P., Gujrat, Bihar, U.P., West Bengal.
10.	Gypsum	23.90	Rajasthan, Tamil Nadu, Jammu and Kashmir, Gujrat
11.	Lime stone	7645	M.P., Tamil Nadu, Andhra Pradesh, Bihar, Gujrat, Rajasthan, Karnataka
12.	Graphite		Kerala, Bihar, Orissa, Rajasthan, Tamil Nadu, Andhra Pradesh

MINERAL RESOURCE MANAGEMENT

World's stock of mineral is considerably depleting because of their use in almost every walk of human activities. Time has come when appropriate attention is to be given to understand the ill effects of reduced mineral supply for the well being of the human beings. There is an urgent need to call for the proper exploration and development and management of mineral of resources in India. The important one are listed as below:

(*i*) Geological Survey of India (GSI)

(*ii*) Indian Bureau of Mines (IBM)

(*iii*) Public Sector Mining Undertakings

These are involved in the survey and search of minerals, conservation and development of mineral resources, providing technical assistance to mining industries, functioning as data bank of mines and minerals and advice central and State Govt. on all aspects of mineral industry.

Mining and oceans: ocean mineral resources are found in three areas: sea water, sediments and deposits on the shallow continental shelf and sediments and nodules on the deep ocean floor. Most of the chemical found in sea water occur in such low concentration that recovering them takes more energy and money. Then they are worth. Only magnesium, bromine and sodium chloride are abundant enough to be extracted profitable at current prices with existing technology.

Deposits of minerals along the continental shelf and near shore lines are already significant sources of sand, gravel, phosphates, sulfur, tin, copper, iron, tungsten, silver, titanium, platinum, and even diamond. Offshore wells also supply large amount of oil and natural gas. The deep ocean floor at various sites may be a future source of manganese and other metals. Manganese rich nodules are thought to cover 20% of the world's ocean floors and have been found in large quantities at a few sites.

Classification of minerals: Minerals are generally classified into three groups on the basis of their use, metallic fuels and non-metallic fuels. Metallic minerals or ores, or rocks or mineral deposits are those form which metals are extracted profitably. The important metals are gold, silver, platinum, copper, lead, iron etc. Among the mineral fuels, coal, petroleum and natural gas have occupied the most important position in the mineral kingdom. The non-metallic minerals or industrial minerals have one or more metals, commonly iron or aluminum, as part of their chemical formulae, but the physical characteristics rather than the metal content determine the use of these minerals.

A mineral for materials has been a human occupation for as long as we have recorded history. Mining for significant energy resources such as coal and uranium is likely to continue for the foreseeable future. Much of the energy resources are mined form the surface, once referred as strip mining, these surface methods are now known as surface mining. Surface mine, open pit mines, and quarries differ in important ways. Surface mines are developed to mine a material which lies a narrow bed relatively close to the surface. The bed of material can be removed and the surface restored such that, if done properly, the site resemble what it was prior to construction. Open pit mines are constructed to mine deep sources of ores which lie diffusely through rock strata. The material is removed and refined, but typically only tailing remains. Open pit mines are thus, very difficult to reclaim. Quarries are developed to mine stone and thus leaving no opportunity to reclaim.

Minerals and environment: The extraction of minerals from the surface of the earth curst is associate with several environmental hazards. *e.g.* coal mining is a dangerous and dirty activity. Under ground mining can cause the collapse of surface areas, while strip mining leaves land scapes scared and useless. Many wastes produced during the process are highly acidic and can contaminated streams and rives if not contaminated. In 1977, the US government enacted the surface mining control act, while requires mining companies to restore land to its original contours, replace top soil, replant vegetation and manage waste water.

Various activities like cutting lanes and hazels, establishing pits, galleries and shafts blasting, dumping excavated material, drilling etc. cause tremendous detrimental effects on environmental health. The important consequences of underground mining and surface mining activities are damage to ecosystem, including air pollution, noise nuisance, alternation of surroundings topography, damage to flora and fauna and social problem related to exploitation conflicts and resettlement.

Dressing of minerals which includes various steps to process the raw materials into marketable products causes various effects on the environment in totality. Following are the environmental risks form the dressing of minerals:

Domain	Risks	Causes
Earth surface / soil	Consumption	Infrastructural measures (traffic, structures) build up of dups, barren rocks, storage and thickening of suspensions and sludges.
General vicinity	Noise, vibrations, dust size silting	Energy generation, transport, handling
Air	Solid contaminants, gaseous and vaporous contaminants, exhaust smoke, fumes	Dust form size reduction, dry screening and handling gas evolution in ponds and dressing facilities
Surface water	Solid tailing, liquid tailing	From burning dumps, coking, roasting, siltation, airborne erosion from dumps, leaching of dumps, improper handling, storage and transportation of operating media
Human	Health, social conflicts	Various forms of use environment
Fauna	Disturbaces/ expulsion poisoning	Noise, air and water pollution chemical agents

Food is the primary need of all the living beings including the human. For centuries, the general rule for basic foodstuffs grains, vegetable, meat and dairy products was self sufficiency. Whenever climate, blight or war interrupted the agricultural production of a region or a nation, the inevitable result was famine and death, sometimes on the scale of millions. With industrial revolution trade between nations intensified and soon it became economically feasible to ship basic foodstuffs from one part of world to another. Today, agricultural production systems do much more than supply a country's internal food needs. For some nations, the capacity to produce more basic food stuffs than the home production needs represents an extremely important entry into international market. And for many developing countries special commodities such as coffee, fruit, sugar, spices, coca and nuts provide their only significant export material.

In past fifty years, the population of the worlds has increased tremendously but food production has not been able to cope up with this increase as a major part of the land which was required for agricultural purposes was acquired for construction of homes and industries. Thus, the shortage of land and less developed techniques of agriculture, lead to shortage of food for the growing population. In India more than 300 million people are victims of mainutrition. The major problem is only because of high population growth. The period 1960-1980 has been one of the best periods for the country in term of food production which is known as "Golden Era" or "Green revolution". The main objevtive of this revolution was make India self sufficient in term of food production. During this period the average yields of corps per hectare go almost double. This made India from food importing country to food exporting country. But still with growing techniques and higher crop yields, many part of the world are facing the problem of food deficiency.

Many country are not self sufficient in food. The U.S. Canada, Australia and nations of European community have been the sources of most of the donated food. However, according to world watch institute, international food aid was cut half between 1993 and 1996 resulting only 7.6 million tons available for distribution. This reflects economic restraints in donar countries as well as weakening political support for such aid. Absolute reliable figures on the world wide extent of hunger are unavailable. The world health organization has estimated that 840 million people are unfed and undernourished. Some 18% populations of the developing world, the region most seriously affected are southern Asia (specially Bangladesh), Latin Americal and Sub-Sharan Africa.

The root cause of hunger is poverty. Poverty means the lack of sufficient income in cash to meet the most basic biological needs for clothing and shelter. A number of Asian countries including China, Indonesia and Thailand significantly reduced the extract of poverty and hunger during 1980's. In china, food security is a matter of high national priority. Indonesia and Thailand have benefited form oil exports and tourism sector respectively. Drought is another reason of famine in the Sharan countries like Sahel, Chud, Sudan, Ethiopia and Somalia.

CHANGES CAUSED BY MODERN AGRICULTURAL PRACTICES

Until 150 years ago, the majority of people of United States lived worked on small farms. Farmers used traditional approaches to combat pests and soil erosion crops were rotated regularly, many different crops were grown and animal wastes were returned to soil regularly. Then in mid 1800's the industrial revolution contributed to a revolution in agriculture so profound that today less than 35% of the W.S. work force producer enough food for all the nations needs plus substantial amounts for trade on world markets.

Before 1960's much of increased production came from new land into production. Any singnificant expansion of cropland will come at the expense of forest and wetlands, which are both economically and ecologically fragile.

Overgrazing: grass lands that receive too little rainfall to support cultivated crops have been used for grazing livestock. Such as lands are too often overgrazed. As grass production fails to keep up with consumption the land becomes barren; wind and water erosion follows and desertification results.

Fertilizers and pesticides: When fertilizers were first employed, 15 to 20 additional tons of grain were grained for each ton of fertilizer used. When levels of fertilizers are too high, plants became more vulnerable to attack by insect and other pests. When fertilizers washed away, the result is water pollution.

Chemical pesticides have provided significant control over insect and pests, but the pests have resistant to most of the pesticides as a result of natural section. Lots of chemicals in use have not been adequately tested in term of human concerns and their potential to cause genetic defects.

Irrigation and water logging: world wide irrigated acreage increased about 26 times from 1950 to 1995. it is still expending, but at much slower pace because limits of water resources. More ominous much present irrigation is not sustainable. In addition, production is being adversely affected on as much as 1/3 of world's irrigated land because of water logging and accumulation of salts in soil, consequences of irrigation where there is poor drainage.

Salinity: salinization is the accumulation of salts in and on the soil to the point that plant growth is suppressed. Approximately 2.5 to 3.75 million acres are salinized each year Salinization occurs because even the fresh water contains at least 200-250 ppm of dissolved salts. As the applied water leaves by evaporation of transpiration, the salts in solution remain behind and gradually accumulate. Because it happens in dry lands salinization considered a form of desertification.

ENERGY RESOURCES

Energy today has become a key factor in deciding the product cost at micro level as well as indicating the inflation and the debt burden at macro level so energy is a primary input for almost all economic activities and is therefore vital for improvement in quality life.

The 90 percent of the world's current energy supply is based on fossil or mineral sources, oil, coal, gas and uranium. Only 10 percent is from renewable sources, especially hydro-power and biomass, while the share of solar and wind energy is below 01 percent. The following table indicates the pattern of primary consumption in the world:

Energy sources	World-wide consumption	Percentage
Oil	21676.10^6 Barrel	34.2
Coal	4765.10^6 tons	30.2
Gas	1923.10^6 cbm	19.1
Nuclear	167 TWA	5.0
Hydro-power	2050 TWA	6.1
Wood/biomass	1219.10^6	5.3
Wind	3 TWA	1.10^{-2}
Solar	0.1 TWH	1.10^{-3}
Geo-thermal energy	76 TWH	0.1

Source: Survey of energy resources, World energy conference, 1989

The energy resources can be classified in many ways:

(*a*) **Commercial fuels:** These include coal, lignite, petroleum products, natural gas and electricity.

Non-commercial fuels: These include fuel wood, cow-dung, agriculture wastes etc.

(*b*) **Primary energy resources:** Primary energy resources are those resources which are mined or otherwise obtained from the environment. These include fossil fuels (coal, lignite, crude oil and natural gas), nuclear fuels, water (hydro energy), solar energy, wind energy, ocean and geothermal energy resources.

Secondary energy resources: Secondary energy resources are those which do not occur in nature; instead, they are derived form primary energy resources. These include petrol, diesel, electric energy (from coal, diesel and gas) etc.

(*c*) **Conventional energy resources:** These include fossil fuels (coal, petroleum and natural gas), water (hydro poser).

Non-conventional energy resources: These includes solar, wind, geothermal, ocean (thermal, tidal and wave), biomass and hydrogen energy.

(*d*) **Non- renewable energy resources:** Non-renewable energy resources are those resources which are exhaustible and cannot be replaced once by they are used. These are available in limited amount and develop over a long period. These include fossil fuels (such as coal, oil and natural gas), and nuclear power.

Renewable energy resources: Renewable energy resources are those natural resources which are inexhaustible (*i.e.* which can be replaced as we use them) and can be used to produce energy again and again. These are available in unlimited amount in nature and develop in a relatively short period of time. These include solar, wind, water, geothermal, ocean and biomass energy. Nuclear energy, however, can also be considered as inexhaustible source of energy if atomic minerals are used in fast breeder reactor technology.

Solar energy: The sun is essentially a thermo nuclear reactor and is almost an inexhaustible source of energy. Sun's heat energy is used directly by plants which are primary producers and this converted into chemical energy through the process of photosynthesis, which enters and utilized by secondary and tertiary producers in the food chain. Sun's heat energy is being used directly in solar cookers to cook food or in solar heater to heat water. Thus sun energy can be concentrated by big concave reflectors and used to boil water to obtain steam and produce electricity.

Solar energy is kinetic energy radiation. Since using solar energy will not deplete the sun in any way. Astronomers estimates that sun will burn for another several billion years, it is therefore commonly referred as renewable energy.

SOLAR RADIATIONS AND ITS SPECTRAL CHARACTERISTICS

The wave concept of solar radiations explains how the energy propagates, but this can only be detected when it interacts with matter. In the interaction these radiations behaves as though it consists of many individual bodies called photons.

Frequency(n): It is the number of wave crests passing a given point in a specified point of time. Frequency was formerly expressed as cycles per second, today it is expressed in terms of hertz, the unit for a frequency of one cycle per second.

Wavelength (l): Wavelength is the distance from any point on one cycle or wave to the same position on the next cycle or wave. The electromagnetic spectrum ranges from the very short wavelengths of the gama rays to the long wave lengths of the radio region.

The following table shows spectral regions of the solar radiations:

Gama rays	less than 0.03 nm
X-ray	0.03 to 3.0 nm
UV rays	0.03 to 0.4 mm
Photographic UV band	0.3 to 0.4 mm
Visible	0.4 to o.7 mm

Infra red	0.7 to 100 mm
Reflected IR band	0.7 to 3.0 mm
Thermal IR band	3 to 5 mm and 8 to 14 mm
Microwave	0.1 to 30 cm
Radar	0.1 to 30 cm
Radio	> 30 cm.

The greatest amount of output of energy is in the visible light part of the spectrum.

The photovoltaic cell or solar cell: Solar cell, technically known as photovoltaic cell, is the device which enables a direct conversion of light energy to electrical energy. Generally constructed from a sandwich of two thin layers of silica which are treated such that light striking one causes electrons to fall to other. Solar energy can be converted directly into electrical energy by photovoltaic (PV) cell, commonly called solar cells. Sun light falling on a solar cell- a transparent water, thinner than a sheet of paper-releases a flow of electrons, creating an electrical current. Because a single solar cell produces only a very small amount of electricity, many cells are wired together in a panel providing 30-100 watts. Several panels are in turn wired together and mounted on a roof and produce electricity for a home or building. The DC electricity produced can be stored in batteries and used directly and converted to conventional AC electricity.

Solar cells are reliable and quite, have no moving parts and should last for 30 years or more if incased in glass or plastic. The cells can be installed quickly and easily and expended or moved as needed; maintenance consists of occasional washing to keep dirt from blocking the sun rays. Arrays of cells can be located in deserts and marginal lands, alongside interstate highways, in yards and on roof tops. In 1995, Enron, a major natural gas company, announced plans for a solar photovoltaic power plant backed up by energy efficient turbines burning natural gas that was expected to generate 100 megawatts of electricity

OCEAN THERMAL ENERGY CONVERSION (OTEC)

In oceans, a thermal gradient (*i.e.* temperature difference) of about 20 °C exists between surface water heated by the sun and colder deep water. This temperature difference can be harnessed to produce power. This concept is known as ocean thermal energy conversion (OTEC). An OTEC power plant can be built on a barrage (*i.e.* sailing vessel) that could travel anywhere in the ocean. It uses the warm surface water to heat and vaporize a low boiling liquid such as ammonia. The increased pressure of the vaporize a low boiling liquid would drive turbo-generators. The ammonia vapors leaving the turbines would then be condensed by cold deep water which is about 100 m below the surface and is returned back to start the cycle again.

Owing to the small temperature differences between the surface water and deep water, the conversion efficiency is as low as 2-3%. This low efficiency by itself is immaterial since the primary energy source viz. the temperature differences between the surface and deep water over most of ocean, is freely available. However, this low efficiently coupled with other drawbacks such as high capital costs, persistent maintenance problems and fouling of pipes and pumps due to marine organisms, results in meager energy yields thereby rendering OTEC power uneconomical at the present state of the OTEC technology.

GEOTHERMAL ENERGY

The energy released in the decay of naturally occurring radioactive materials, the earth interior is huge ball molten rock, which occasionally erupts to the surface in volcanoes. This heat of the earth's interior is referred to as geothermal energy. It can be obtained at practical coast; geothermal energy is virtually infinite and ever lasting. Some geothermal energy may be obtained where natural water comes in contact with hot rocks. Such heated water may come to the surface in natural steam vents. The steam is used to drive turbogenerators. As of 1988, there were about 5000 megawatts of such geothermal facilities installed, about half of them in California and other Philippines, Mexico, Italy and Japan. This amount could be increased by several folds in the coming years.

Large scale of development of geothermal power presents many problems, hot stream and water brought to the surface are frequently heavily laden with salt and other contaminants, particularly sulfur compounds. These contaminating compounds are highly corrosive to turbines and other equipments and they result in both in air and water pollution as they are finally released into the environment. Sulfur pollution from a geothermal plant may be equivalent to that from a plant burning high sulfur coal, and the hot brines released in to stream and rivers may be ecologically disastrous.

The real potential for geothermal energy lies in development of hot dry rock. Hot dry rock can be found any where by drilling deep down. In theory, two parallel holes cab be drilled in to hot rock and then fractures created between the two holes. Water forced down one hole will be heated as it seeps through the fractures and will come up the other holes as steam to drive turbine. Recycling the water would the problem of pollution.

TIDAL POWER

A great deal of power energy is inherent in the twice daily rise and fall of the tides, and many imaginative schemes have been proposed for capturing this external, pollution free sources of energy. The most straightforward is to build a dam across the mouth of way and mount turbine in the structure. The incoming tides flowing through the turbine generates power. As the tide shifts, the blades are reversed so the out flowing water continues to generate power. Two such plants are currently in operation, one in France and the other in the Russia. But tidal power is not without adverse environmental impact. In addition to loss of unique aesthetic and recreational pleasures in these areas, there would be far reaching environmental effects due to the dam's trapping the sediments, impending of the migration of marine organism, and from changing circulation and the mixing of fresh and salt water.

HYDRO-ELECTRIC POWER

Water energy is most conventional renewable energy sources and obtained from water flow, water falling form a height. Hydro-electric power plants convert the kinetic energy contained in falling water into electricity. Water power has been used since ancient times by diverting water from natural streams or rivers over various kinds of paddle wheels or turbines. The power output form waterwheels being low, people started buildings dams form the last centaury to obtain a substantial head of hydrostatic pressure. Thus water under high pressure, flows through the base of the dam and drives turbo-generators producing hydroelectric power.

Hydro-electric power not only a clean, non-polluting source of energy, but also a most conventional renewable energy resource. Further, with the limited sources of coal, lignite and oil, growing reliance would have to be placed on hydel power. Except for the heavy initial investment, hydel projects have a definite edge over other power plants. Hydro power is one of the most important sources of energy, next only to thermal power. This can be assessed by the fact that nearly 30 % of the total power of the world is met by hydro-electric power. The total hydro potential of the world is about 5,000 GW. There are countries in the world where almost entire power generation is hydro based.

At present about 24% of the India's electricity is being generated through the hydro power plants. Out of the total installed power generating capacity of 92,864 MW (at the end of 1998-99) in the country, 22,438.48 MW is from hydel plants. As per estimates of central electricity authority, the annual hydroelectric potential of our country at 60% load factor is 89,830 MW; yet hardly 25% of it has been harnessed so far. About 80 percent of the developed hydel resources of India lie in the states of Maharashtra, Tamil Nady, Karnataka, Kerala, Punjab, Himachal Pradesh, Jammu and Kashmir and Uttaranchal.

WIND ENERGY

It is an inexhaustible energy resource which was used in very old times in grinding grains, lifting water and propelling ships. The instrument that converts wind energy into electrical energy is called wind mills. Netherlands is land of wind mills. A steady wind of more than 25 km/hr is required to run a wind mill, e.g. coasts, mountains, certain valleys and plains. A number of wind mills can be installed at one plac. Wind rotates as wheel connected to a generator or turbine. Asia's largest wind mill complex is installed at Lamba (Gujrat). In India, wind energy production was only 2 MW in 1989 and 950 MW in 1998.

Wind power is a non-polluting, renewable and hence sustainable source of energy. However, it has the following drawbacks:

(*i*) Location of wind farms on migratory routes could spell hazard to birds and disastrous form some avian populations.

(*ii*) Their appearance on the landscape and their continual whirring and whistling can be irritating.

BIOMASS BASED ENERGY

Biomass is the term used for all materials originating form photosynthesis. Thus biomass includes all new plant growth, residues and waste (wood, short rotation trees) hebaceous plants, fresh water and marine algae aquatic plants, agriculture and forest residue (straw husks, bagases, corn cobs, bark, sawdust, wood shavings, roots, animal dropping), wastes (night soil, sewage, garbage), biodegradable organic effluents form industries like tanneries, sugar mills, slaughter houses, meat packing plants breweries distilleries. Biomass energy systems are renewable and sink for CO_2 helps to conserve soil and water and helps also in water run off and in halting desertification march.

Petro-plants: There are attempts to identify potential plant species as source of liquid hydrocarbons, a substitute for liquid fuels. The hydrocarbon present in such plants can be converted into petroleum hydrocarbons. Several species of plants belonging to families like Euphorbiaceae, Asclepidaceae, Anacardiaceae and Labiatae are considered as most suitable energy plants. *Euphorbia lathyris* (Graphar) has more than 5% oil and polymeric hydrocarbons. Some other examples of petroplants are *Calotropos procera, C. giganatea, Euphorbia royleana* and recently discovered *Jojoba curcas* and *Euphorbia antisyphilitica.*

Biogas: Biogas is an important solution to present energy crisis, especially in rural areas. This is an environmentally clean technology. About 75% of the dung would be available for biogas. All India average for dung production is 11.30 kg for cattle and 11.60 for buffalos with 67.10 m^3 gas per ton of wet dung. Biogas is composed of Methane, CO_2, H_2 and N_2.

In order to make best use of biogas technology two things are relevant restricted use of water and better strains of methane generating bacterial for normal microbial activity 90% water content is needed. Thus we need additional water which may be critical to this technology in dry areas of the country. There is to be developed a dry process that requires less of water. Also there is need for methnogens, able to operate at temperature of lower than 20 °C or more. Biogas can also be generated from and one such plant is operation at Okhla, Delhi. There have been developed Urja Grams by using biogass, animal, poultry and human excreta.

Dendrothermal energy: Denuded wastelands are being used for plantation of fast growing shrubs and trees with high calorific value. They in turn provide fuel wood, charcoal, fodder, power and also scope for rural employment. Through gasification system these energy plantation over, 8,000 ha were producing nearly 1.5 MW power in 1987.

Fossil fuels: They are nonrenewable conventional energy resources found inside earth's crust where they have been formed through heat and compression on forests and other organic matter buried underneath due to earthquakes, landslides, lava etc. Since they were formed in particular periods, their presence underneath can be known from specific palynofossils/ microfossils. Fossils fuels can be solid (coal, lignite), liquid (petroleum) or gaseous (natural gas). They meet 70% of total energy needs of the world and 87.4% of all commercial energy. Per capita index of energy consumption is MTOE (metric tones of oil equivalent). It is maximum for Canada (9.15 MTOE), high for U.S.A. (7.3 MTOE), mediocre for Britain and France (3.8 MTOE) and low for India (0.5 MTOE). In India 58% of commercial energy is got from coal and 38% from petroleum along with natural gas. Coal is used for cooking, heating, in industry and thermal power plants. Petroleum is used for transport, agricultural and some industries. LPG is liquefied petroleum gas. Natural gas is used for employed both in cooking and industry. Fossil fuel resources are, however, limited. Coal resources and natural gas may last for over 100 years. Good coal reserves occur in U.S.A, china, Russia and India (Bihar, Bengal, Orissa and M.P.). In future, it may be liquefied petroleum gas. A lot of natural gas goes waste (some 24% in India). Known reserves of petroleum are expected to last upto 2030 AD. Major reserves occur in Nigeria, Libya, Iran, Iraq, South Arabia, Kuwait, U.S.A, Russia and China. In India, petroleum reserves are low being mainly present in Assam, Gujarat, Bombay High and Narmada Basin. They do not meet even 50% of total petroleum requirement. Despite limited reserves, demand for fossil fuels is rising, annually by 6 % on global basis and 15 % in India. Therefore, conservation of fossil fuels,fprevention of their wastage and alternfate sources of energy are urgently required. Gasification of coal, compressed natural gas (CNG), gasohol (petro + alcohol) are being tried.